COMMUNICATION WITH EXTRATERRESTRIAL INTELLIGENCE

Other Pergamon Titles of Interest

CURTIN
Trends in Communications Satellites

GRAYBIEL
Basic Environmental Problems of Man in Space

NAPOLITANO
Space Activity, Impact on Science and Technology
Space Stations, Present and Future
Space and Energy
A New Era in Space Transportation
Using Space Today and Tomorrow, Volumes 1 & 2

WRIGLEY
Space Relativity

Pergamon Journals of Related Interest

Acta Astronautica

Planetary and Space Science

Progress in Aerospace Sciences

Vertica

COMMUNICATION
WITH
EXTRATERRESTRIAL INTELLIGENCE

Editors

JOHN BILLINGHAM

and

RUDOLF PEŠEK

NASA, Ames Research Center, Moffett Field, CA 94035, U.S.A.

PERGAMON PRESS

OXFORD · NEW YORK · TORONTO · SYDNEY · PARIS · FRANKFURT

U.K.	Pergamon Press Ltd., Headington Hill Hall, Oxford OX3 0BW, England
U.S.A.	Pergamon Press Inc., Maxwell House, Fairview Park, Elmsford, New York 10523, U.S.A.
CANADA	Pergamon of Canada, Suite 104, 150 Consumers Road, Willowdale, Ontario M2J 1P9, Canada
AUSTRALIA	Pergamon Press (Aust.) Pty. Ltd., P.O. Box 544, Potts Point, N.S.W. 2011, Australia
FRANCE	Pergamon Press SARL, 24 rue des Ecoles, 75240 Paris, Cedex 05, France
FEDERAL REPUBLIC OF GERMANY	Pergamon Press GmbH, 6242 Kronberg-Taunus, Pferdstrasse 1, Federal Republic of Germany

First edition 1979

British Library Cataloguing in Publication Data

Communication with extraterrestrial intelligence

1. Life on other planets. 2. Interstellar communication

I. Billingham, John II. Pešek, Rudolf

III. 'Acta astronautica'

001.5'0999 QB54 79-40048

ISBN 0 08 024727 X

Published as a Special Issue of the journal *Acta Astronautica,* Vol. 6, Nos 1-2, and supplied to subscribers as part of their subscription. Also available to non-subscribers.

Printed in Great Britain by A. Wheaton & Co. Ltd., Exeter

CONTENTS

Acta Astronautica. Vol. 6, pp. 1–2. Pergamon Press 1979. Printed in Great Britain

Preface

THIS SPECIAL ISSUE of *Acta Astronautica* is devoted to the subject of Communication with Extraterrestrial Intelligence (CETI).

The appearance of this volume is a reflection of a steadily growing interest in the possibility that intelligent species may be widely distributed in the Universe. This hypothesis is based on major advances over the past twenty to thirty years in the sciences of astrophysics, astronomy, planetary formation, exobiology and biological evolution. To make a long story short, it is now believed that planets are the rule rather than the exception, that life will arise in suitable planetary environments, and that in many cases life will evolve to the stage of intelligence, given several billion years of comparative stability of the planetary environment.

The crucial question is immediately apparent. Is there any way in which we can establish communication with other civilizations? The probability of the existence of a second intelligent species in our own solar system is vanishingly small. Hence we are driven to examine possibilities for interstellar communication. It transpires that advances in communication theory and technology over the last decade or two do indeed offer some reasonable approaches to the formidable problems involved.

Thus the development of a scientific basis for the evolution of intelligence is proceeding in parallel with the emergence of technologies which allow our first attempts at interstellar communication. In both areas the challenge has been seized by a small but growing number of investigators of widely differing disciplinary backgrounds. The papers in this volume are the results of a rich variety of approaches to the challenge.

CETI was introduced to the International Academy of Astronautics by Rudolf Pešek of Czechoslovakia in 1966. The First International Review Session on CETI took place in 1972 as a part of the XXIIIrd Annual Meeting of the International Astronautical Federation. In subsequent years the CETI Review Session has been a regular feature of the program. In 1977 the Program Committee approved the introduction of a second CETI Review Session in response to increasing interest in the subject.

This volume contains most of the papers presented at these Review Sessions in 1975, 1976 and 1977. It contains also a paper on a design concept for a large space radiotelescope, by V. I. Buyakas and several other Soviet authors from the Space Research Institute, USSR Academy of Sciences, which was presented in another session at the 1977 IAF meeting. We have introduced one further paper. It is the first part of a recent publication entitled "The Search for Extraterrestrial Intelligence: SETI"† prepared for the NASA Ames Research

†The term CETI has been used for many years as a generic title for all aspects of Communication with Extraterrestrial Intelligence. The term SETI was introduced in the United States in 1976 to reflect the widely held view that the primary endeavor should be a *Search* for Extraterrestrial Intelligence.

Center by a team of U.S. Scientists under the chairmanship of Philip Morrison. We feel it is of interest because it represents an up-to-date condensation of many of the fundamental questions and conclusions about SETI at this time.

On the whole, the papers in this volume deal with concepts and studies related to the science, technology, and observational techniques of CETI. Indeed, they include prominent papers by Soviet authors dealing with searches that have actually been carried out. One of the papers, by Pešek, is a summary of major CETI activities in science, technology, and other areas during the years 1966 through 1976. Finally, we have included one or two papers dealing with some very advanced concepts of communication.

We consider it an honor to have been involved in the publication of this Special Issue of *Acta Astronautica*. We hope that this volume will serve as a useful store of knowledge about some of the CETI activities in recent years. And we trust it will stimulate debate on the complex and important issues involved for years to come.

We must acknowledge those who have worked so hard to accomplish the publication of these collected papers. Particular thanks go to Charles Seeger of San Francisco State University for his generous assistance in editing the papers; to Ms. Beulah Gossett of the NASA Ames Research Center for her impecable work in the preparation of the manuscripts; to Ms. Kathy Michaud for her excellent typing; to all the people of the Graphics Branch at Ames for their skills in refining the figures for the text; to Dr. James Lawless of Ames for editorial assistance; to Dr. A. K. Oppenheim, Editor-in-Chief of *Acta Astronautica* for his guidance and patience; and last, but not least, to those very important people, the authors.

JOHN BILLINGHAM
RUDOLF PEŠEK

Acta Astronautica. Vol. 6, pp. 3–9. Pergamon Press 1979. Printed in Great Britain

Activities of the IAA CETI committee from 1965–1976 and CETI outlook

R. PEŠEK
Czechoslovak Academy of Sciences, Prague, Czechoslovakia

Abstract—On 10 October 1966, the scientists entrusted by the International Academy of Astronautics (IAA) to discuss CETI problems met for the first time in Madrid. After 10 years, it is interesting to recall the development of CETI ideas and to think about the outlook. Since the spring of 1965, IAA has organized an International CETI Symposium. For this purpose, a Study Group was established in September 1965, followed by an Organizing Committee in April 1966 and a CETI Standing Committee in September 1974. For different reasons, the Symposium was postponed several times and eventually cancelled. Half-day CETI Review Meetings were then held every year starting in 1972. The following studies and projects concerning CETI via radio or via starship were elaborated: Project Cyclops (U.S.A.), Program CETI (U.S.S.R.), Project Daedalus (British Interplanetary Society), National Space Program for Interstellar Exploration (Hughes Research Laboratories, U.S.A.) and Outlook for Space (NASA). The outlook for the IAA CETI Committee might be defined in this way: to continue in organizing CETI Review Meetings, to invite scientists to suggest experiments concerning CETI, to coordinate CETI national programs, to incite the interest of the United Nations in CETI and to help them in all CETI problems, and to have all important decisions concerning CETI considered by the United Nations.

1. Introduction

SIXTEEN YEARS have passed since the International Academy of Astronautics was established at Stockholm under the guidance of the late Theodore von Kármán. One of the aims of this Academy is to study projects for space exploration and investigations that would benefit greatly from close international cooperation.

Communication with extraterrestrial intelligence (CETI) is, by its very nature, international in character both as regards its possible accomplishment and its implication for all mankind. Since 1965 the IAA has been concerned with CETI. Ten years ago on 10 October 1966, the first session of the CETI Study Group took place in Madrid. Ten years is not a long time for projects like Apollo or Concord, 10 years for CETI is a milestone. Now, when the Viking probes are searching for life on Mars, it would perhaps be interesting to sum up the history of CETI at the International Academy of Astronautics and to estimate the CETI outlook.

2. CETI in the IAA

How did the CETI problem appear to be in the spring of 1965? Now and then, some lectures on CETI were delivered at the IAF Congresses by A. G. Haley, H. Strughold, A. E. Slater, J. Gadomski and R. Margaria.

The first attempt to detect possible extraterrestrial radio signals was carried out in the United States in 1960—Project Ozma.

In November 1961, an informal conference was held at the Green Bank Observatory in West Virginia. Its task was to discuss whether it might be possible to contact other worlds. The conference was not held in secret in the official sense, but every effort was made to avoid publicity because of the sensational nature of the question to be discussed (Sullivan, 1965).

In May 1964, the Academy of Sciences of the Armenian SSR organized a first All-Union Conference on Extraterrestrial Civilizations and Interstellar Communication. A resolution recommending further theoretical and experimental CETI research was approved.

In the spring of 1965, Mariner 4 was on its journey to Mars; on 14 July 1965, the spacecraft passed within 9800 km of the Martian surface and 21 pictures were transmitted to Earth.

These were all rather national endeavors; I have thought and I still think that a search for extraterrestrial civilizations should represent efforts of all mankind. That is why the idea to organize an *international* symposium on CETI started. For this symposium, I devised the word CETI—an acronym for Communication with Extraterrestrial Intelligence. The choice is also connected with the well-known fact that Ceti in Latin is the genitive of Cetus (whale) and that Tau Ceti is a star 11.8 light years distant, rather similar in size and luminosity to our Sun. With some probability, Tau Ceti has a planetary system.

On 15 May 1965, at the seventh session of the IAA Board of Trustees in Paris, I proposed CETI as the subject of a three-day IAA Symposium. The problems suggested for discussion concerned astronomy, biology, communications, and linguistics; the impact of CETI on mankind might also be considered. The Board requested me to approach a limited number of distinguished scientists for their personal opinions as to whether the matter was mature enough to be discussed in a year or two and to report back to the Board of Trustees at its session in Athens.

On 20 July 1965, according to this request, I posted letters to 50 scientists requesting their opinions on a CETI Symposium. The conclusions from this inquiry were:

26 replies (52%) reached Prague in time
16 scientists (61.5%) supported the idea
6 scientists (23.1%) did not think the subject was mature enough at that time
4 replies (15.4%) were rather difficult to judge.

It is interesting to compare the conclusions of this inquiry with the results of another inquiry arranged by the Czechoslovak bi-monthly *RADAR* in June 1966. Of 38 participants, 28 (74%) expressed the opinion that there is a very high probability of the existence of other civilizations; 19 (50%) believed that attempts to establish communication would be worthwhile; 3 participants (8%) did not believe that other civilizations exist, and 7 (18%) replies were rather difficult to judge.

On 15 September 1965, I reported at the Eighth Session of the Board of Trustees in Athens on the replies I had received. In view of the encouraging response, the Board approved the project and recommended that a Study Group be set up to further consider the program and planning of a CETI Symposium and to report to the Board at its spring session. The IAA President appointed a Study Group with the following members:

C. J. Clemedson (Sweden)
V. L. Ginzburg (U.S.S.R.)
A. G. Haley (U.S.A.)
E. B. Konecci (U.S.A.)
R. Pešek, Chairman (Czechoslovakia)
J. S. Shklovskii (U.S.S.R.)
F. L. Whipple (U.S.A.).

Two members were later added: S. M. Beresford and F. I. Ordway.

The only meeting of this Study Group took place in Madrid on 10 October 1966. The Study Group submitted to the President of the IAA the following recommendations:

To hold a 4- to 5-day Symposium which would bring together individuals who are interested in this matter. Such a symposium would be the first international meeting on the subject. The following problems were suggested for discussion: (1) Astronomical problems; (2) Biological problems; (3) Problems of communication; (4) Problems of linguistics; (5) Basic laws of development of civilizations; (6) Impact of CETI on mankind.

It was recommended that a series of proposals be made to interested organizations such as IAU and UNO in this project and that a committee should be set up to organize the meeting sponsored by the Academy; meanwhile, efforts should be made to find support for the meeting.

In April 1967 at its Eleventh Session in Paris, the Board of Trustees authorized the President to proceed with plans to hold a meeting on CETI in 1968 and approved the President's proposal to appoint an Organizing Committee, including the following members:

C. J. Clemedson (Sweden)
F. D. Drake (U.S.A.)
E. B. Konecci (U.S.A.)
B. Lowell (Great Britain)
P. M. Morrison (U.S.A.)
F. I. Ordway (U.S.A.)
R. Pešek, Chairman (Czechoslovakia)
C. Sagan (U.S.A.)
J. S. Shklovskii (U.S.S.R.)
M. Subotowicz (Poland).

In August 1967, the General Assembly of the IAU met in Prague; there I met N. C. Kardashev, J. S. Shklovskii and C. Sagan. Among other things, we proposed the list of participants at the CETI Symposium.

On 27 September 1967, the first meeting of the CETI Organizing Committee took place during the 17th International Astronautical Congress in Belgrade. Since that time, the members of the Committee have met regularly at all the International Astronautical Congresses and the IAA Board of Trustees has discussed CETI problems during all its sessions.

On 25 April 1968, the Presidium of the Czechoslovak Academy of Sciences authorized the CETI symposium in Czechoslovakia and agreed to meet the cost of lecture rooms and of interpretation.

On 18 October 1968, the meeting of the CETI Organizing Committee was held in New York and the Committee agreed on a tentative schedule of topics for the CETI Symposium and recommended that the Symposium be held in Prague, Czechoslovakia, 7–12 September 1969. A tentative list of 55 participants was prepared. The following program was recommended: (1) Introduction to the Meeting; Astronomical Perspective; (2) Origin of Life and Life in the Solar System; (3) Evolution of Intelligence; (4) Evolution of Technical Societies; (5) Potential Sites for Extraterrestrial Intelligence; Search Problems; (6) Signal Acquisition; (7) CETI and Mankind; (8) Prospects and Recommendations for Future Research. For different reasons, we have postponed the Symposium several times.

In the meantime, however, the events led in another direction. Independently, the Astrosoviet of the Academy of Sciences of U.S.S.R. organized a CETI international conference. To realize it with participants from many countries appeared rather complicated. Therefore, N. S. Kardashev and C. Sagan proposed to have a Soviet–American conference first, with a small number of participants. This conference, sponsored jointly by the National Academy of Sciences of the United States and the Academy of Sciences of the Union of Soviet Socialistic Republics was held at Byurakan, U.S.S.R., from 5–11 September 1971.

On 19 September 1971, the CETI Organizing Committee met in Brussels. I reported there on the Byurakan Conference and proposed to cancel the CETI Symposium in Prague and organize a half-day International CETI Review Meeting at the next International Astronautical Congress in Vienna. The Board of Trustees agreed.

Since this decision, review meetings were held in Vienna (1972), Baku (1973), Amsterdam (1974), Lisbon (1975), and this year (1976) at Anaheim, California (U.S.A.). A list of lectures delivered at these meetings is in the Appendix.

In view of the growing interest in CETI and theoretical and experimental studies being pursued, especially in the U.S.S.R. and in the United States, the Board of Trustees decided to set up a Standing Committee on CETI to replace the former Organizing Committee. On 10 September 1974, the IAA President invited the following scientists to serve as members of the new Committee:

J. Billingham (U.S.A.)
C. J. Clemedson (Sweden)
V. V. Gogosov (U.S.S.R.)
A. T. Lawton (United Kingdom)
G. Marx (Hungary)
R. Pešek, Chairman (Czechoslovakia)
C. Ponnamperuma (U.S.A.)
M. Subotowicz (Poland).

Representatives of other countries should be invited to support us in our endeavor.

3. CETI outside the IAA

How did the CETI problems appear to be in the summer of 1976? Since 1965 they had been influenced by many events on Earth and in space. I shall recall some of them:

(1) New simple organic compounds detected by radioastronomers in interstellar space and the finding and analysis of the Murchinson meteorite have increased our confidence that life could evolve on other worlds (Ponnamperuma, 1972).

(2) The search for signals from space, using existing radiotelescopes, is under way in the U.S.S.R., United States and Canada. These attempts have not been successful thus far.

(3) Pioneers 10 and 11, the first vehicles that will leave our solar system, carry a plaque, informing from where, when, and by what kind of beings these vehicles were launched.

(4) On 16 November 1974, the first deliberate attempt to send a radio message was made. The message was transmitted from the 300-m Arecibo radiotelescope in Puerto Rico and was aimed at the star cluster Messier 13.

(5) Soviet and American space vehicles photographed the surface of Mars. Two Viking probes were launched on 20 August and 9 September 1975, toward Mars; their most important aim is to search for extraterrestrial life.

(6) Five Soviet Venera probes landed on Venus and returned data and photographs from there. They found conditions not favorable for life.

(7) A series of lectures on CETI problems was organized in the United States, U.S.S.R., Poland, German Democratic Republic and Czechoslovakia. On 23 September 1974, an exhibit "Life in the Universe" was opened at the Smithsonian National Air and Space Museum in Washington, D.C.

(8) Many informative books on CETI were published. More and more papers on CETI were printed in scientific reviews. Very extensive bibliographies on extraterrestrial life and CETI were published in the United States and in the U.S.S.R.

4. CETI outlook

Some comprehensive CETI studies and projects were elaborated in the United States, the U.S.S.R. and the United Kingdom. The most important seem to be:

(1) Project Cyclops—a Design Study of a System for Detecting Extraterrestrial Intelligent Life. In the summer of 1971, this study was performed by a team of scientists at the NASA–Ames Research Center. The concept is a large array of phased radiotelescopes on Earth, listening in the microwave region of the spectrum for coherent signals from star systems within 1000 light years.

(2) In March 1974, the Board of the Scientific Council of the Radio Astronomy Area of the Soviet Academy of Sciences considered and approved a Research Program on Communication with Extraterrestrial Civilization. In this program, two instrumentation projects, CETI 1 (1975–85) and CETI 2 (1980–90), are included. The periodical Astronomicheskii Zhurnal published this program in Russian, and the reviews in *Soviet Astronomy* (1975) and *Journal of the British Interplanetary Society* (1975) in English. I think that this program might be used as a basis for the preparation of international CETI programs.

In these studies of contact with extraterrestrial intelligence, attention was focussed on communication via radio signals. Other studies are focussed on interstellar flights.

(1) Project Daedalus is the name chosen for the British Interplanetary Society's Starship Study. An attempt is made to design an unmanned probe to fly near Barnard's star without the necessity of involving radically new concepts or of extrapolating present-day capabilities too far into the future (Bond and Martin, 1975).

(2) "A National Space Program for Interstellar Exploration" is the title of R. L. Forward's study (1975) concerning interstellar flights, unmanned and manned. This study assumes that around the year 2000 a number of automated interstellar probes will be launched, and in the year 2025 a manned exploratory starship will be launched.

A slightly different work is NASA's *Outlook for Space* (Hearth, 1976). "The purpose of this study is to examine future space activities within the context of possible national needs and to identify directions the United States should take in the civilian use and exploration of space for the remainder of this century." The study has 12 themes. The last theme is called "Origins and Future of Life" and includes the following objectives: (1) How did life on Earth originate? (2) Is there extraterrestrial life in the solar system? (3) What organic chemistry occurs in the universe? (4) Do other stars have planets? (5) Can we detect extraterrestrial life?

5. Conclusion

From these *national* studies and projects, this outlook for the IAA CETI Standing Committee might be proposed:

(1) To continue organizing international CETI review meetings;

(2) To invite scientists to suggest experiments concerning CETI for radio-telescopes, satellites, interplanetary probes, Spacelab, and space telescope;

(3) To coordinate CETI national programs;

(4) To interest the United Nations in CETI and help them in all CETI problems;

(5) To endeavor—by all means we may have—to establish the rule and make it a practice that all really important decisions concerning CETI are to be taken by the United Nations.

References

Bond A. and Martin A. R. (1975) Project Daedalus. The origin and aims of the study. *J. Br. Interplanetary Soc.* **28**, 147–149.

The CETI Program (1975) *Sov. Astron.* **18**(5), 669–675; translated from *Astron. Zh.* **51**, Sept.–Oct. 1974, pp. 1125–1132.

Forward R. L. (1975) *A national space program for interstellar exploration.* Hughes Research Laboratories, Malibu, Calif., Research Rept. 492.

Hearth D. P. (Study Director) (1976) *Outlook for Space.* Report to the NASA Administrator by the Outlook for Space Study Group. NASA SP-386.

Ponnamperuma C. (1972) *The Origins of Life.* Thames and Hudson, London.

Sullivan W. (1965) *We Are Not Alone.* Hodder and Stoughton, London.

USSR Research Program on the Problem of Communication with Extraterrestrial Civilizations (1975) *J. Br. Interplanetary Soc.* **28**, 751–760.

Appendix

Summary of IAA International CETI Review Meetings

First meeting, 14 October 1972, in Vienna, Austria:

(1) Survey of CETI Activities up to Present—R. Pešek, Czechoslovak Academy of Sciences, Prague, Czechoslovakia.

(2) Astronomical Aspects of CETI—S. von Hoerner, National Radio Astronomy Observatory, Green Bank, West Virginia, U.S.A.

(3) State of the Art in the Detection of Intelligent Extraterrestrial Signals—B. M. Oliver, Hewlett-Packard Company, Palo Alto, Calif. U.S.A.

(4) Methods of Deciphering a Message from Extraterrestrial Intelligence—B. V. Sukhotin, Institute of the Russian Language, U.S.S.R. Academy of Sciences, Moscow, U.S.S.R.

(5) State of the Art of Biological Problems—L. M. Muchin, Laboratory of Cosmic Biology, U.S.S.R. Academy of Sciences, Moscow, U.S.S.R.

Second meeting, 13 October 1973, in Baku, U.S.S.R.:

(1) The Search for the Radio Emission from Extraterrestrial Civilizations—V. S. Troitskii *et al.*, Radiophysical Research Institute, Gorky, U.S.S.R.

(2) Search for Signals from Extraterrestrial Civilizations by the Method of the Synchronous Dispersion Reception—L. M. Gindilis *et al.*, Shternberg Astronomical Institute, Moscow, U.S.S.R.

(3) On Some Estimates Characterizing Energetically Optimal Trajectories of Interstellar Monitor Probes—U. N. Zakirov, Institute of Applied Mathematics, U.S.S.R. Academy of Sciences, Moscow, U.S.S.R.

Third meeting, 4 October 1974, in Amsterdam, the Netherlands:

(1) Survey of CETI Activities in the Years 1971–1974—R. Pešek, Czechoslovak Academy of Sciences, Prague, Czechoslovakia.

(2) Interpreting Signals from an Interstellar Probe—A. T. Lawton, Shepperton, Mddx., United Kingdom.

(3) About Some Maneuvers of Subrelativistic Probes—U. N. Zakirov, Institute of Applied Mathematics, U.S.S.R. Academy of Sciences, Moscow, U.S.S.R.

(4) Interstellar Communication by Neutrino Beams—M. Subotowicz, University of M. Curie-Shklodowska, Lublin, Poland.

Fourth meeting, 27 September 1975, in Lisbon, Portugal:

(1) The Latest Events in CETI—R. Pešek, Czechoslovak Academy of Sciences, Prague, Czechoslovakia.

(2) A Review of the Theory of Interstellar Communication—J. Billingham *et al.*, NASA–Ames Research Center, Moffett Field, Calif. U.S.A.

(3) Some Energetic Estimation of a Relativistic Spacecraft Ensemble—U. N. Zakirov, Institute of Applied Mathematics, U.S.S.R. Academy of Sciences, Moscow, U.S.S.R.

(4) Corrosion-Free Interstellar Probes—G. Marx, Roland Eötvös University, Budapest, Hungary.

(5) On Some Newly Defined Problems of CETI—M. Subotowicz, University of M. Curie-Shklodowska, Lublin, Poland.

Acta Astronautica. Vol. 6, pp. 11–31. Pergamon Press 1979. Printed in Great Britain

The search for extraterrestrial intelligence—SETI

PHILIP MORRISON

Massachusetts Institute of Technology, Cambridge, MA 02139, U.S.A.

JOHN BILLINGHAM AND JOHN WOLFE

NASA, Ames Research Center, Moffett Field, CA 94035, U.S.A.

Abstract—This article is taken from a NASA Special Publication, SP-419, released in March 1978. The publication deals with most of the current major topics in the field of the Search for Extraterrestrial Intelligence (SETI), and, as described in the Preface, was the result of a two year series of Science Workshops carried out for NASA by a team of eminent U.S. scientists and engineers under the chairmanship of Dr. Philip Morrison. It seemed to us to be of sufficient interest to extract the first part of the SETI publication SP-419 and include it in this issue of Acta Astronautica. We have therefore taken the first 35 pages of SP-419 and reproduced them in the following pages. They embrace the CONSENSUS of the Science Workshop Team members, and their CONCLUSIONS. Not included here is the rest of SP-419 which gives more detailed coverage to the science and technology of SETI in Part II, Colloquies, and Part III, Complementary Documents. Those wishing to have a complete copy of SP-419 should write to Dr. John Billingham, NASA Ames Research Center, Moffett Field, CA 94035, U.S.A.

NASA Special Publication (SP-419)

TABLE OF CONTENTS

MEMBERS

Science Workshops on Interstellar Communication

Philip Morrison, Chairman	Massachusetts Institute of Technology
Ronald Bracewell	Stanford University
Harrison Brown	California Institute of Technology
A. G. W. Cameron	Harvard University
Frank Drake	Cornell University
Jesse Greenstein	California Institute of Technology
Fred Haddock	University of Michigan
George Herbig	University of California–Santa Cruz
Arthur Kantrowitz	AVCO Everett Research Laboratory
Kenneth Kellermann	NRAO, Green Bank
Joshua Lederberg	Stanford University
John Lewis	Massachusetts Institute of Technology
Bruce Murray	Jet Propulsion Laboratory
Bernard Oliver	Hewlett–Packard
Carl Sagan	Cornell University
Charles Townes	University of California–Berkeley

Preface

Over the past two decades there has developed an increasingly serious debate about the existence of extraterrestrial intelligent life. More recently, there have been significant deliberations about ways in which extraterrestrial intelligence might in fact be detected. In the past two years, a series of Science Workshops has examined both questions in more detail. The Workshop activities were part of a feasibility study on the Search for Extraterrestrial Intelligence (SETI) conducted by the NASA Ames Research Center.

The objectives of the Science Workshops, as agreed at the second meeting in April 1975, were: to examine systematically the validity of the fundamental criteria and axioms associated with a program to detect extraterrestrial intelligent life; to identify areas of research in the astronomical sciences, and in other fields, that would improve the confidence levels of current probability estimates relevant to SETI; to enumerate the reasons for undertaking a search, the values and risks of success, and the consequences of failure; to explore alternative methods of conducting a search; to select, in a systematic way, preferred approaches; to indicate the conceptual design of a minimum useful system as required to implement the preferred approaches; to delineate the new opportunities for astronomical research provided by the system and their implications for system design; to outline the scale and timing of the search and the resources required to carry it out; to examine the impact of conducting a search, and the impact of success or failure in terms of national, international, social and environmental considerations; and to recommend a course of action, including specific near-term activities.

This report presents the findings of a series of Workshops. The major conclusions of our deliberations are presented in Section I. First, an Introduction lays out the background and rationale for a SETI program, and then in The Impact of SETI, we examine the implications of the program. In particular, the Impact section examines the significance of the detection of signals and of information that may be contained in signals from extraterrestrial civilizations.

For those who wish to see some of the arguments in more detail, we have extracted from the discussions of the last two years, six of the most interesting and significant elements of the debate in the forms of Colloquies. Finally, we have documented, in greater depth, a selection of detailed technical arguments about various aspects of the SETI endeavor.

The reader should note that the Introduction, the Impact of SETI, and the Conclusions, which comprise Section I of this volume, have been prepared by and represent the views of the Workshop as a whole. Sections II and III, on the other hand, have been prepared by the individual authors listed, and while consonant with the major SETI findings, reflect specifically the views and style of presentation of the authors.

In addition to the series of six Workshops, and at the instigation of the participants, two additional series of meetings were held. The first, under the Chairmanship of Dr. Joshua Lederberg of Stanford University, addressed the question of Cultural Evolution in the context of SETI. The second, under Dr. Jesse Greenstein of the California Institute of Technology, addressed the question of the Detection of Other Planetary Systems. The conclusions of these meetings are presented in Colloquies 2 and 3.

The last of the Complementary Documents (II–15) lists the members of the Science Workshops, our consultants and advisors, and the agendas for the nine Workshop meetings. Detailed minutes of all of the Workshops are available from Dr. John Billingham, SETI Program Office, NASA Ames Research Center, Moffett Field, CA 94035, U.S.A.

I would like to express my appreciation to everyone who has worked with me in this undertaking. I must single out first the Workshop members themselves (see Complementary Document 15), and in particular Joshua Lederberg and Jesse Greenstein for their major contributions in taking the chair at their respective special Workshops (see Colloquies 2 and 3). The assistance of the NASA Centers, and specifically of the SETI Groups at the Ames Research Center and Jet Propulsion Laboratory must be recognized, together with numerous contributions from consultants and speakers who have addressed and advised us. Last, but by no means least, special thanks are due to Vera Buescher, Secretary to the Ames SETI Team, for her loyal and indefatigable attention to the thousand details which went into the preparation of this report.

In conclusion, I would hope that our report will provide a logical basis for the evolution of a thoroughgoing but measured endeavor that could become a significant milestone in the history of our civilization.

We recommend the initiation of a SETI program now.

PHILIP MORRISON
Chairman

Section I: CONSENSUS

Introduction

> Heaven and Earth are large, yet in the whole of space they are but as a small grain of rice.... It is as if the whole of empty space were a tree, and heaven and Earth were one of its fruits. Empty space is like a kingdom, and heaven and Earth no more than a single individual person in that kingdom. Upon one tree there are many fruits, and in one kingdom many people. How unreasonable it would be to suppose that besides the heaven and Earth which we can see there are no other heavens and no other earths?
>
> TENG MU, 13th Century philosopher
> (Translated by Joseph Needham)

In the enormous emptiness of space we can now recognize so many stars that we could count one hundred billion of them for each human being alive. Yet we know of only one inhabited planet, our Earth. The Earth has supported the development of life nurtured by one commonplace star, the nearby five-billion-year old Sun. We look out into the starry Universe quite unable to see within its compass any sign that we are not alone. The other planets near our Sun offer some hope to a search for other life, and indeed for many months Viking on the surface of Mars has been reporting the enigmatic chemical activity of the Martian soil. We remain uncertain, at the time of writing, whether the chemical changes are biological or inorganic in nature.

The web of life here on Earth is the consequence of a long complex sequence of natural selection by which life increased its scope and its variety, always exploiting the flood of energy bestowed directly or indirectly by the Sun. The Earth has seen fire and ice, yet it has provided steadily, for three billion years without a break, some environments to which life could adapt. Changes were never so drastic or so rapid that all survival became impossible, though particular species have arisen and died by the millions. Indeed, life has spread from its origins, probably near the seashore, to alpine peaks and ocean troughs, and has diversified almost beyond description. Our species and a few of our forebears have achieved considerable technological abilities and some degree of self-knowledge. Nor do we foresee any natural catastrophe ending this fabric of life until in due course the Sun itself runs out of nuclear fuel, some five billion years in the future.

We all know the starry sky at night, and on our deep photographs of the sky we see everywhere a dusting of small dots. Analysis of the light which caused those images, using its intensity and the details of its spectrum, has made it certain that such dots represent suns resembling our own, about which we know only that they are suns. Our own Sun with its cortege of planets would be just such another single dot, quite indistinguishable from a hundred million others at the distances we scan.

We have been able to understand in a general way how stars are born out of dense clouds of gas and dust in the interstellar spaces; we can see other stars in the transient stages of birth, as once was the Sun and its planets. Are planets always born in the spinning disk of gas out of which the other suns form? Or is our own set of planets as rare as its central luminary is commonplace? We cannot now say, though we are sure that the processes that form stars and might have formed planets as well, were going on for billions of years before our solar system formed, and will outlast our Sun.

If around those other visible suns there spin other planets, hidden from us by the distances of space, it is at least possible that on some the work of natural selection has continued for times which could be five or ten billion years longer than the whole history of our Sun and Earth. We could conceive that life never arose on a given planet, or that it exhausted its resources of adaptability, to end in an algal monotony, or in total extinction. Or we can imagine the slow evolution of beings—not human—able to control their world and themselves and to know the Universe.

In evolutionary diversity there is still unity. Squid and human see with eyes that evolved quite differently, and yet resemble each other closely in streamlined form, and even somewhat in behavior. They are distinct evolutionary solutions to the problem of earning a living by predation upon fast-swimming fish; the three, fish, reptile, and mammal, have been molded alike by natural selection to solve the single dynamical problem of fast pursuit in the sea. Similarly, the way of life of *H. sapiens* appears to spread and to succeed; it seems to us that if natural selection has once built so subtle and successful a scheme, it can do so again. Sapient beings on other planets would in no way be our biological kin for they would share with us no common ancestor. But they might have converged with us in behavior; they might have evolved to culture, and then, say, to radio telescopes. Culture is a workable way of life, like hunting schools of mackerel. Indeed, we have seen that human cultural evolution, also, often converges: no less a development than writing was independently achieved by the Aztecs, the Chinese, and the peoples of the Middle East. On this basis, it would be consistent with the historical development of the great ideas of science to postulate that for a time of unknown duration, near an unknown number of stars, deliberate radio beacons or unintended radio leakage are present. This is a hypothesis untested, but capable of verification by experiment.

It is not idle curiosity to inquire whether other intelligent life has arisen and survived near some distant sun, beings in no way our biological kin. For by some sign of that presence we might come to learn, in a way, our own possible future. Indeed, the one most solid result of the calculus of chance which governs our thoughts about such uncertainty is this: intelligent beings out there—if they exist at all—almost surely form societies which have endured for a time long compared to the history of our own civilization, a time which might even reach the span of geological time itself.

Astronomers have real hope of detecting planets near other stars, especially relatively neighboring ones, by new optical and infrared measurements from ground or orbit. But detection of plant or animal life implies a landing such as we made on Mars, and this is well beyond our capabilities over interstellar distances.

If we are to learn about distant life, it must make itself perceptible. As far as we can see, only life that has followed our own evolution to the extent of being able to send *some* mark of its presence across space can be found. This must mean that intelligence develops naturally out of evolving life, that it can make signals capable of traversing space, and that, for some period of time at least, it wants to make its presence known (or at least does not conceal it!). If these conditions exist anywhere, we might hope to detect creatures far older and more capable than ourselves. Exploration would then cross a new frontier, the frontier of an intelligence biologically wholly unrelated to our own.

How would such signals be made? Might super-Viking probes cross space? Might light flashes like stellar lighthouses show an intelligent presence? Much speculation has considered the situation (some of the variety of different ideas are presented in Complementary Document 1). The key facts seem to be that radio waves cross space well, and that the radio engineer has found means to detect extremely weak signals with large dishes and extremely sensitive receivers. Violent events on every scale, from explosions in galaxies to electrical instabilities on the planet Jupiter, have been recorded by radio astronomers. None of these signals appear to bear the marks of any but an astronomical origin, so far. Interesting as these have been, it remains true that radio energy compared to visible light is scarce in the Galaxy. Within the scope of present knowledge in our own Galaxy, a certain well-defined radio waveband (from about one meter to one centimeter) is, for natural reasons, the quietest region over the whole span of electromagnetic waves (see Section II–4).

This fact lies behind a remarkable event in human history. Almost imperceptibly, without really intending it, within the past two or three decades we have entered a new communicative epoch. Until that time, we could have made no sound, no pattern or mark, no explosive flash of light on our small planet that could be detected far out among the stars by any means we understand. Space is too deep, and the stars are rivals too brilliant, for any mere faint human glow to become visible far away. Even the whole amount of sunlight reflected from a planet, a light source thousands of times more powerful than all the energy now at human disposal—is still beyond our ability to pick out at the distance of a nearby star. But our radio technique, only a generation or so old, has now reached such maturity that a signal sent from an existing radio dish on Earth, with sending and receiving devices already at hand, could be detected with ease across the Galaxy by a similar dish, if only it is pointed in the right direction at the right time, tuned to the right frequency. Such a lucky observer—or one who is patiently and systematically searching—would see us as unique, distinguished among all the stars, a strange source of coherent radio emission unprecedented in the Galaxy.

Or are we without precedent? Are we the first and only?

Or are there in fact somewhere among the hundred billion stars of the Galaxy other such beams, perhaps so many of them that our civilization, like our Sun, is to be counted as but one member of a numerous natural class? For such a radio beam cannot come, we think, from any glowing sphere of gas or drifting beam of particles. It can come only from something like our own complex artificial apparatus, far different from any star or planet, smaller, newer, much more particular; something we would recognize as the product of other understanding and ingenious beings.

That is the topic of this technical report: the search for extraterrestrial intelligence, SETI.

We do not intend to send any signals out to add to those which have already gone out from our TV transmitters and our powerful radars. Rather, we want to listen, to search all the directions of space, the many channels of the radio (and other) domains, to seek possible signals. Perhaps it will be only an accidental signal, as we have made ourselves. That would be harder to find. Or perhaps there is a deliberate signal, a beacon for identification, or even a network of communication. There seems no way to know without trying the search. This is an exploration of a new kind, an exploration we think both as uncertain and as full of meaning as any that human beings have ever undertaken.

The search would be an expression of man's natural exploratory drive. The time is at hand when we can begin it in earnest. How far and hard we will need to look before we find a signal, or before we become at last convinced that our nature is rare in the Universe, we cannot now know.

The impact of SETI

Whether the search for extraterrestrial intelligence succeeds or fails, its consequences will be extraordinary.

If we make a long dedicated search that fails, we will not have wasted our time. We will have developed important technology, with applications to many other aspects of our own civilization. We will surely have added greatly to our knowledge of the physical Universe. The global organization of a search for interstellar radio messages, quite apart from its outcome, can have a cohesive and constructive influence upon our view of the human condition. But above all, we will have strengthened belief in the near uniqueness of our species, our civilization and our planet. Lacking any detection, the conviction of our uniqueness would hardly ever reach certainty; it would form over a long time, less into sharp conclusions than into a kind of substructure of human thought, a ruling consensus of attitudes. If intelligent, technological life is rare or absent elsewhere, we will have learned how precious is our human culture, how unique our biological heritage, painstakingly evolved over three or four thousand million years of tortuous evolutionary history. Even a growing possibility of such a finding will stress, as perhaps nothing else can, our lonely responsibilities to the human dangers of our time.

On the other hand, were we to locate but a single extraterrestrial signal, we would know immediately one great truth: that it is possible for a civilization to maintain an advanced technological state and *not* destroy itself. We might even learn that life and intelligence pervade the Universe. The sharpness of the impact of simple detection will depend on the circumstances of discovery. If we were to find real signals after only a few years of a modest search, there is little doubt the news would be sensational. If, on the other hand, signals were detected only after a protracted effort over generations with a large search system, the result might be less conspicuous.

Note well that it is likely that the early announcements of the detection of deliberate signals may turn out to be mistaken, not verified by further study and observation. They may be natural phenomena of a new kind, or some terrestrial signal, or even a hoax. (Indeed, this has already happened—more than once!) Press and public must use caution if we are to escape the volatile raising and dashing of great hopes. We stress the importance of a skeptical stance and *the need for verification*, because we hold that even a single genuine detection would in and of itself have enormous importance.

Of course it is very difficult to foresee the content of a signal except in the most general way. A signal could be a beacon—a deliberate transmission specifically for the purpose of attracting the attention of an emerging civilization like ourselves. Alternately, it could be a leakage signal similar to our own television broadcasts or radars, not intended for our detection. Whatever the signal, we would remind the reader that it will be a one-way transmission. Any message in such a transmission would be a message between cultures not between persons. We have human analogies at hand, in our long-continued interest in great books from the past, say the Greek philosophers; we ponder them afresh in each generation, without any hope of interrogating Socrates or arguing with Aristotle.

The first authentic signals will attract intense headline attention. But after that the pace must slow. Perhaps we will learn only that the signal exists. This alone will be significant. We will know we are not alone. However, the information content of any signal could be rich. Study would continue for decades, even generations. Books and universities will be more suited for the news than the daily programs. If the signal is deliberate, decoding will be relatively easy, we expect, because the signal will be anticryptographic; made to reveal its own language coding. If the message comes by radio, both transmitting and receiving civilizations will have in common at least the details of radiophysics. (The commonality of mathematics and the physical sciences is the reason that many scientists expect the messages from extraterrestrial civilizations to be decodable—if in a slow and halting manner.) No one is wise enough to predict in detail what the consequences of such a decoding will be, because no one is wise enough to understand beforehand what the nature of the message will be.

Some have worried that a message from an advanced society might make us lose faith in our own, might deprive us of the initiative to make new discoveries if it seems that there are others who have made those discoveries already, or might have other negative consequences. But we point out that we are free to ignore an interstellar message if we find it offensive. Few of us have rejected schools because teachers and textbooks exhibit learning of which we were so far ignorant. If we receive a message, we are under no obligation to reply.† If we do not choose to respond, there is no way for the transmitting civilization to determine that its message was received and understood on the tiny distant planet Earth. (Even a sweet siren song would be little risk, for we are bound by bonds of distance and time much more securely than was Ulysses tied to the mast.) The receipt and translation of a radio message from the depths of space seem to pose few dangers to mankind; instead it holds promise of philosophical and perhaps practical benefits for all of humanity.

Other imaginative and enthusiastic speculators foresee big technological gains, hints and leads of extraordinary value. They imagine too all sorts of scientific results, ranging from a valid picture of the past and the future of the Universe through theories of the fundamental particles to whole new biologies. Some conjecture that we might hear from near-immortals the views of distant and venerable thinkers on the deepest values of conscious beings and their societies! Perhaps we will forever become linked with a chain of rich cultures, a vast galactic network. Who can say?

If it is true that such signals might give us, so to speak, a view of one future for human history, they would take on even greater importance. Judging that importance lies quite outside the competence of the members of this committee, chosen mainly from natural scientists and engineers. We sought some advice from a group of persons trained in history and the evolution of culture, but it is plain that such broad issues of the human future go beyond what any small committee can usefully outline in a few days. The question deserves rather the serious and prolonged attention of many professionals from a wide range of disciplines—anthropologists, artists, lawyers, politicians, philosophers, theologians—even more than that, the concern of all thoughtful persons, whether specialists or not. We must, all of us, consider the outcome of the search. That search, we believe, is feasible; its outcome is truly important, either way. Dare we begin? For us who write here that question has step by step become instead: *Dare we delay*?

---

†It is for this reason that this undertaking is not called *Communication* with Extraterrestrial Intelligence (CETI), but *Search* for Extraterrestrial Intelligence (SETI).

First conclusion

It Is Both Timely and Feasible to Begin a Serious Search for Extraterrestrial Intelligence

Only a few decades ago most astronomers believed that planetary systems were extremely rare, that the solar system and the habitat for life that Earth provides might well be unique in the Galaxy. At the same time so little was known about the chemical basis for the origin of life that this recent event appeared to many to verge on the miraculous. No serious program for detecting extraterrestrial intelligence (ETI) could arise in such an intellectual climate. Since then numerous advances in a number of apparently diverse sciences have eroded the reasons for expecting planetary systems and biogenesis on suitable planets to be unlikely. Indeed, theory today suggests that planetary systems may be the rule around solar type stars, and that the Universe, far from being barren, may be teeming with life, much of it highly evolved. (See Section II–1 and II–3.)

During the latter half of the last and the first part of this century, the slow rotation of the Sun stood as a formidable objection to the nebular hypothesis of Kant and Laplace, which proposed that planetary systems formed out of the same condensing cloud that produced the primary star. An initial rotation rapid enough to produce the Sun's planets should have produced a Sun spinning a thousand times faster—too fast to become a spherical star. As a result, various "catastrophic" theories of the origin of the solar system were proposed, all of which depended on events so rare as to make the solar system virtually unique.

Then, in the late 1930's, Spitzer showed that starstuff torn out by tidal or concussive forces would explode into space rather than condense into planets. Shortly thereafter research into plasma physics, and observations of solar prominences, revealed the magnetohydrodynamic coupling of ionized matter to magnetic fields, a mechanism whereby stars in the process of formation can slow their rotation. As a result, the theory in which planets condense out of the whirling lens of gas and dust that will become a star has regained wide acceptance. Planetary systems are now believed to exist around a substantial fraction of stars. (See Section II–3.)

Meanwhile the discoveries that the organic building blocks for DNA and proteins can be formed by natural processes out of molecules comprising the early atmosphere of Earth, and that many organic molecules are even formed in the depths of interstellar space, have made the spontaneous origin of life on suitable planets seem far more probable. Life appears to have developed on Earth almost as soon as seas had formed and chemical evolution had provided the building blocks. Earth has been lifeless for only a small fraction of its age. This leads many exobiologists today to look upon life as a very likely development, given a suitable planet. (See Section II–1.)

The present climate of belief makes it timely to consider a search for extraterrestrial life, but is such a search feasible? It is certainly out of the question, at our present level of technology or, indeed, at any level we can foresee, to mount an interstellar search by spaceship. On the other hand, we believe it is feasible to begin a search for signals radiated by other civilizations having technologies at least as advanced as ours. We can expect, with considerable confidence, that such signals will consist of electromagnetic waves; no other known particle approaches the photon in ease of generation, direction and detection. None flies faster, none has less energy and is therefore cheaper than the radio frequency photon. It has long been argued that signals of extraterrestrial origin will be most apt to be detected in the so-called microwave window: wavelengths from about 0.5 to 30 cm. Natural noise sources rise to great height on either side of this window, making it the quietest part of the spectrum for everyone in the Galaxy. We concur with these arguments. (See Section II–4).

Existing radio telescopes are capable of receiving signals from our interstellar neighbors, if of high power or if beamed at us by similar telescopes used as transmitters. The large antenna at Arecibo could detect its counterpart thousands of light years away. Indeed, it could detect transmissions from nearby stars less powerful but similar to our own television and radars.

Terrestrial UHF and microwave emanations now fill a sphere some twenty light years in radius. This unintended announcement of our technological prowess is growing stronger each year and is expanding into space at the speed of light. The same phenomenon may well denote the presence of any technological society. In fact, our own radar leakage may have already been detected by a nearby civilization. In addition, advanced societies may radiate beacons for a variety of reasons, possibly merely to bring emerging societies into contact with a long established intelligent community of advanced societies throughout the Galaxy. A search begun today could detect signals of either type.

We propose a search for signals in the microwave part of the radio spectrum, but not at this time the sending of signals. Even though we expect our society to continue to radiate TV and radar signals we do not propose to increase our detectability by, say, intentionally beaming signals at likely stars. There is an immediate payoff if we received a signal; transmission requires that we wait out the round trip light time before we can hope for any results. Transmission should be considered only in response to a received signal or after a prolonged listening program has failed to detect any signals. (See Section II–5.)

Not only is the technology for discovering ETI already at hand, but every passing year will see the radio frequency interference (RFI) problem grow worse while only modest improvements in technology can occur. (See Sections III–8 and III–9.) Perfect receivers would not double the sensitivity of a search system over that which we can already achieve. Given optimum data processing, large increases in sensitivity are to be had only by increasing collecting area. It is true that data processing technology is improving rapidly, but presently achievable data processing technology is adequate and inexpensive. Further, the techniques need to be developed in association with existing facilities and comprehensive searches made before it becomes evident that a more sensitive system is needed. Great discoveries are often the result more of courage and determination than the ultimate in equipment. The Niña, the Pinta, and the Santa Maria were not jet airliners, but they did the job.

Second conclusion

A Significant SETI Program With Substantial Potential Secondary Benefits Can Be Undertaken With Only Modest Resources

A large, expensive system is not now needed for SETI. If we but equip existing radio telescopes with low-cost state-of-the-art receiving and data processing devices, we will have both the sensitivity to explore the vicinity of nearby stars for transmitters similar to Earth's, and to explore the entire Galaxy for more powerful signals, or for signals beamed at us. Such explorations, even should they yield negative results, would decrease our uncertainty concerning whether intelligent life transmitting powerful signals may lie beyond our solar system. At the very least, it would be of great interest and some importance either to know we have near neighbors, or to be reasonably confident no nearby transmitting civilizations exist. If, *after* we have made such modest searches, it seems important to us to embark upon a more ambitious SETI program, such as contemplated by the Cyclops study, the experience we will have gained will prove not only invaluable, but essential. Moreover, we expect to derive spin-off benefits of no small significance.

SETI hardware

The arguments for electromagnetic waves as the communications medium seem compelling. The case for the microwave window seems very strong. The reasons for preferring the low end of the window are also strong, but not so strong that higher frequencies in the window should be ignored. The "water hole" between the H and OH lines is an especially attractive band that may be ideal for long range beacons. (See Sections II–4 and III–1.)

ETI signals, particularly those intended for detection by other searching societies, will probably be narrow in bandwidth compared with natural sources and may have monochromatic components which are as narrow as the interstellar medium permits. This increases their detectability at a given radiated power and distinguishes them from the natural background. The hardware needed for SETI therefore consists of an antenna or antenna system, low-noise wide-band receivers to cover the low-frequency end of the microwave window, means of resolving the received spectrum to a very high degree and means to search out and identify automatically any spectral anomalies.

Since halving the system noise temperature is equivalent to doubling the system sensitivity, it is important in SETI to have the lowest noise receivers that can be built. The background temperature in the preferred frequency region is only 6–8 K (3–5 K in space) so every degree of reduction in receiver noise temperature is significant. The development of suitable low noise receivers represents a simple extension of present microwave technology and is not an expensive program. It would also benefit deep space communications and radio astronomy. (See Section III–5.)

To search for narrow band signals that may be anywhere in a wide frequency band and to do so in a reasonable time has been one of the major challenges of a SETI. In the Cyclops system concept the received signal was optically transformed into a high-resolution power spectrum. Since 1971 the growth of large-scale integrated circuit technology has been spectacular. It now appears possible to build, at reasonable cost, solid state fast Fourier analyzers capable of resolving the instantaneous bandwidth into at least a million channels on a real time basis. Development of such equipment is again a modest undertaking and the equipment would be very valuable for many other uses besides SETI. (For example see Section III–5.)

To complete the data processing it is necessary to examine the power spectrum or a succession of samples of the power spectrum for any sort of significant pattern such as a sustained peak that may drift slowly in frequency, a regularly recurring peak, or arrays of regularly spaced peaks, to name but a few. The data rates are so great that this pattern recognition must be automated. The principal problems associated with the pattern recognition system are the amount of data storage needed and the identification of the types of patterns to be recognized. Only a few years ago these could have presented severe difficulties, but the solid state electronics revolution has so reduced the cost of memory, that prospective data processing costs appear to be relatively inexpensive.

It has been estimated that the development of the right data processing equipment would increase the capability of existing radio telescopes to detect ETI signals by about a thousandfold. This means that very significant searches can be made using existing antennas so equipped and it is recommended that the search begin in this way. The possibility of discovering some unknown type of natural source in this way must not be overlooked.

Search strategies

It is not feasible to search for all kinds of signals at all frequencies from all directions to the lowest flux levels at which a known signal of known frequency and direction of arrival can be detected. (See Sections II–5 and III–2.) The more inclusive the search becomes in frequency or spatial direction, the more time is required, unless we sacrifice sensitivity. This is, of course, the reason for making use of all available *a priori* information and guesses as to preferred frequencies and likely directions of arrival. Many ingenious arguments have been offered for special frequencies and directions or even times; all can be given some weight as the search proceeds. On the other hand, every reduction in some dimension of the search is based on an assumption that may be wrong.

The strategy of searching nearby F, G and K main sequence stars at ever increasing range seems very natural; the only life we know lives on a planet around a G2 dwarf star. This strategy takes us only as far into space as necessary to discover our nearest radiative neighbors around such stars. On the other hand, only slightly older cultures may be capable of radiating much more powerful signals, or they may know that life is to be found only around a few stars of a certain spectral class and age and may beam signals at these. As is true for stars, the nearest transmitters may not be the brightest. The strongest signals may come from advanced societies at great distances, whose transmitters may not even be near any stars.

For these reasons it is premature to adopt only one strategy to the exclusion of others. To cover a wide range of other possibilities it is recommended that in addition to a high sensitivity search of nearby stars, there also be a complete search of the sky to as low a flux level and over as wide a frequency band as practicable. (See Sections II–5 and III–3.)

To be significant, a full sky survey should be able to detect coherent radiation at a flux level one or two orders of magnitude below that provided by existing radio astronomy surveys. This turns out to be easier than one might expect. Although a sky survey as sensitive as $\sim 3 \times 10^{-23}$ W/m^2 has been made this has covered only ~2% of the sky. Another, covering most of the sky, has been made to a sensitivity of $\sim 2 \times 10^{-20}$ W/m^2. But in these, as apparently in all radio astronomy sky surveys, any coherent signals that might have been present were rejected as "interference." Thus a complete sky survey using SETI data processing equipment to detect coherent signals at flux levels of $\sim 10^{-20}$ to $\sim 10^{-24}$ W/m^2 would be very significant. Existing antennas could be used to search the water hole to this level and the entire microwave window to as low as $\sim 10^{-23}$ W/m^2 in a few years observing time.

The target search of the nearer F, G and K main sequence stars should be conducted using SETI hardware with existing antennas. This would permit detection of coherent signals at a flux level as low as $\sim 10^{-27}$ W/m^2, or 10^3 to 10^7 times weaker than for the full sky search, assuming an observation on the order of a half hour per star.

Both the sky survey and the targeted search could produce positive results, but even negative results will be of value since the upper limit flux levels that would result will be much lower than before. This could change our assessment of the capabilities of other intelligent life. The experience gained using SETI hardware in actual operation, with natural and man-made interference present, will affect the design of any future search strategies, and may lead to modifications of hardward, software, and search procedure. The searches we propose can be completed in approximately five years.

Planning a dedicated facility

SETI is more than a single effort. Like the exploration of the New World by our forefathers, like the present exploration of our solar system, it should be accomplished by many missions, each with some particular goal in mind. But there is a limit to the time that can be reasonably devoted to SETI from the facilities of radio astronomy or other services. To achieve the ultimate goals of SETI it will probably be important to have a dedicated SETI facility, the planning for which should begin now. This facility may never need to grow beyond a collecting area equivalent to one, or a few 100-m dishes. That will depend on future priorities, and on what we learn from the searches we immediately propose. The facility may be on the ground, or in space. (See Section III–7.) We should, however, keep possible future needs in mind, and be prepared to build it whenever and wherever it appears appropriate.

Supporting activities

Several ancillary programs should be initiated and pursued. These include protection of the water hole (1.400 to 1.727 GHz) against radio frequency interference (RFI) (see Sections II–4, III–8 and III–9), the detection of extrasolar planetary systems (see Section III–3), the development of techniques for compiling extensive lists of target stars (see Section III–4), the study of alternative search strategies, and the continuing study of the cost effectiveness of space *versus* ground based systems.

In a resolution adopted at its fourth meeting the Science Workshop recommended that international protection of the water hole against RFI be sought at the 1979 World Administrative Radio Conference. (See Section III–9.) Navigational satellite systems are presently being planned that would destroy the usefulness of this prime band of frequencies for SETI purposes. It is important to realize that for ground-based SETI systems such protection does not exclude all other services from the water hole, but only interfering ones such as satellites and nearby ground services. The RFI problem for space based SETI systems (especially systems in synchronous orbit) is more complex and probably more serious. Adequate shielding may be very expensive. It is not necessary that RFI protection of the water hole continue for all time. If no signals are found after a protracted sensitive search, the SETI priority may be relinquished.

The *sine qua non* of SETI is the plenitude of other planetary systems. While theoretical considerations suggest that planetary systems are common, it would be valuable to know how common and how their architecture varies with stellar class and multiplicity. Earlier astrometric telescopes and data reduction techniques could be improved to the point where the existence of near-by planets could be proved or disproved, but the effort might require two or three periods of a major planet, i.e. two or three decades. Preliminary calculations indicate that the direct observation of major planets around nearby stars should be possible with space telescopes of only modest size (on the order of one meter diameter). This could be accomplished by fitting the space telescope with a suitable filter or mask which greatly improves the contrast of a large planet with respect to the central star. Such an approach, if successful, would permit planets to be found in only two to three years after launch. This and other space techniques for direct planetary detection deserve active study and support. (See Section III–3.)

Present star catalogues list the coordinates of F, G and K main sequence stars within only a few tens of light years of the Sun. If we ultimately carry on a search out to several hundred light years we will need to know the location of a thousand times as many target stars as are now listed. The problem of how best to conduct a whole sky star classification and cataloging program needs to be studied and, when solved, to be implemented. Since the compilation of such a target star data base must precede a major search, it is timely to begin the design study now. Both a greatly expanded catalogue of the solar neighborhood and knowledge about nearby planetary systems would be significant contributions to galactic and stellar astronomy as well as to SETI. (See Sections II–6, III–4 and III–6.)

Although it is assumed that the searches performed in this program will be mainly for narrow band signals at the low end of the microwave window, other possibilities should not be ignored. Given a matched filter a series of pulses is just as easy to detect as a continuous wave (CW) signal of the same average power. The pulsed signal, however, introduces the new dimensions of pulse shape, repetition rate, and duty cycle. At the same time it is not clear that CW signals are more probable than pulses. Continuing study of these and other alternatives is indicated.

It will be seen that the program advocated above is of modest scale yet has potential for both SETI success and scientific contribution. Above all it serves as a logical introduction to the future but does not constitute a blank check commitment to a large expensive effort. The program is not a dead end nor is it open ended. It will be timely to consider whether to proceed with a larger scale program after this earlier effort has shown us more accurately what might be involved.

Third conclusion

Large Systems of Great Capability Can Be Built if Needed

Large systems, involving construction of new antennas, are not now needed for SETI. Until we have completed an observational program as suggested in the second conclusion, there seems to be no reason to construct any facility much larger in scale than Arecibo. However, we may some day decide to embark on a more comprehensive search. This could require a system of great capability. Although we emphasize that we do not now recommend construction of such a system, we also feel that it is important to emphasize that a large SETI system is well within the capability of present-day technology.

The first feasibility study of large SETI systems was the 1971 Cyclops project. It concluded that we indeed have the technology to construct a very large ground-based phased array. The system considered would be capable of operating over the 1–10 GHz region of the microwave window, and could grow to collecting areas of many square kilometers if necessary. Its receiving system would be coupled to a data processing system capable of resolving 200 MHz of spectrum into 0.1 Hz channels and of detecting any coherent signal whose power equalled the noise power in this 0.1 Hz bandwidth.

At the request of the Ames Interstellar Communication Study Group, the Jet Propulsion Laboratory performed a detailed independent review of the Project Cyclops report, and found the study to be correct in its major technological conclusions. Today the data reduction would probably use large scale integrated circuit hardware exclusively, rather than optical processors. Today the system noise temperature could be nearer to 10 than 20 K. But these improvements only reinforce the basic conclusion that ground-based systems can be built that will detect a gigawatt omnidirectional beacon or its equivalent at a distance of 1000 light years. This corresponds, in the water hole, to a flux of one photon per second per square kilometer.

The principal cost of the Cyclops system was found to be the antennas. If the effectiveness of the data processing could be improved enough to double the sensitivity for the same antenna area, the original system performance could be achieved at about half the cost. Clearly in systems having large collecting area it is very important to make optimum use of that area by doing the best possible job of data processing. Further studies of the coherent signal detection problem and the possible tradeoffs in time and money *versus* antenna area are needed and should be started now.

Ground-based vs. space systems

Following the Cyclops study the Interstellar Communication Study Group at Ames contracted with the Stanford Research Institute to study various alternatives to a ground-based array in achieving large collecting area. A dozen alternatives were considered, four ground-based, four lunar-based, and four in space. (See Section III–7.)

The study revealed that very large, very lightweight, single unit antennas in space may be cost competitive with a large ground-based array. This conclusion can only be stated as a possibility and not as a fact because of the obvious difficulty of making valid cost comparisons between the well understood, mature ground-based antenna technology and the poorly understood, untested technology of large, lightweight space antennas.

In addition to the primary feasibility and cost of the space structures many other problems associated with space systems need further study. These include the means to shield the receiver against the severe RFI expected in space, the provision for wide band data links on a continuous basis, and the logistics of servicing and maintaining and operating a complex system in space. However, space systems also give unique advantage with respect to system noise, sky and frequency coverage and tracking ability. All that can be said at present is that space systems must be carefully considered in future plans.

Obviously, the whole question of Earth- *versus* space-based systems needs an order of magnitude more study before the issue can be resolved; this must be done before a commitment is made to any large search system. The possibility exists that a combination of ground and space systems would offer advantages not to be found in either alone.

Intermediate steps

As discussed in the second conclusion, a small dedicated facility for SETI will probably eventually be desirable. This will most likely be a single new ground-based, or small space-based, antenna of advanced design, or both. If the facility is ground-based, it would be prudent if its site and design are chosen to ensure that the system be expandable at least to an intermediate size, such as a small array of 6 to 18 antennas. Such a system would increase the sensitivity well beyond that achievable with any existing antenna and would permit simultaneous searches using different strategies. It would also allow phasing techniques to be tested.

With respect to space-borne antennas, it may be desirable, as studies proceed, to fly one or more medium size designs as shuttle payloads. The missions should be designed not only to test the structures but also to allow actual SETI and radio astronomy observations to be made in space. These antennas in conjunction with a dedicated ground facility could be used together as a very long baseline interferometer of greater capability than any now employed in radio astronomy. In addition observations could be made throughout the wide frequency bands over which the atmosphere is noisy or opaque.

Scientific applications

As soon as a dedicated SETI facility achieves either a sensitivity or spectral coverage not found in present radio or radar astronomy instruments, it becomes a uniquely useful tool for research in these areas. An almost continuously increasing spectrum of applications exists as the SETI facility is expanded in scope. It is recommended that a fraction of the time of any dedicated facility be devoted to scientific research which that facility alone makes possible. This might well provide a series of discoveries which in themselves help justify the cost of the SETI facility. (See Sections III–5 and III–6.)

We see that either in space or on the ground the SETI effort can effectively grow from the initial effort to one using a very large system at whatever rate is appropriate. Early studies are needed to refine concepts of large systems, and especially to evaluate the usefulness of space. Even in the absence of the discovery of ETI signals, useful discoveries in science will accrue as the facility expands.

Fourth conclusion

SETI is Intrinsically an International Endeavor in Which the United States can Take a Lead

The Search for Extraterrestrial Intelligence offers benefits for all nations. The search would certainly be facilitated by, and may even require, international cooperation. It is a serious exploration, as important as any ever undertaken, and surely of larger scope than the journeys to the Earth's poles early in the century. We can hope for relatively quick results, but must prudently prepare for a protracted effort. The program must be kept open and public in the spirit of international science and exploration. We can and should expect growing cooperation with investigators from many countries, both those already displaying interest and activity, as the Soviet Union and Canada, and others whose interest would grow.

SETI is not only a response to the spirit of exploration but is natural to the metaphysical view of modern man. The question "Are we alone?" is pertinent to the entire species, both to us and our descendants.

International cooperation is essential to solving the radio frequency interference problem discussed above. Furthermore, it is possible that antennas may be required at various places throughout the world or in space; a system beyond the borders of any single nation. It seems clear to us that the SETI effort should be cast as a cooperative international endeavor at the start and that appropriate international relationships should be established through existing or novel international organizational arrangements. Joint funding is a desirable goal for such an approach. In any case the extended period which may be required for the detection of extraterrestrial intelligence—much less communication—emphasizes the need for organizational and cultural support more enduring than typically characteristic of national programs.

There may be a particular opportunity for joint Soviet and U.S. efforts in the SETI. The Soviets have already begun a preliminary search. Their published discussion of this problem indicates that considerable interest exists within the scientific community there (see Section III–11). The USSR is capable of substantial space technology should that prove important in the future. Finally, joint leadership of an international SETI program by the U.S. and the Soviet Union might constitute a logical continuation of the cooperative endeavors in space initiated by the Keldysh–Low agreements most recently responsible for the Apollo–Soyuz Test Program.

West European nations, especially West Germany, Holland and England, have also evidenced increasing interest in new radio astronomical endeavors. Thus, the possibility of initiating a SETI program through bilateral or multilateral arrangements involving the U.S. warrant consideration as well.

The United States can lead in the SETI endeavor

The United States has frequently demonstrated the will and foresight to take the initiative in programs of worldwide benefit. The U.S. space program has provided not only excitement and scientific knowledge, but numerous practical satellite services not for this country alone, but for the whole world. It is in this same spirit of providing a focal point for international cooperation and support that we feel the U.S. can and should take the initiative in SETI.

The material, technological and intellectual resources of the U.S. are such that a large-scale SETI program could be carried on indefinitely by this country alone without appreciable drain on the economy. There are good reasons for believing the net effect on the economy could be positive. Even if international cooperation and support were slow to materialize, we believe SETI remains a feasible and worthwhile U.S. endeavor.

The psychology of and mechanisms for international cooperation suggest that an international SETI effort is unlikely until one big nation, such as the U.S., seizes the initiative and invites serious participation by others. It is in this sense of initiative and not in the pursuit of narrow national advantage that we recommend a leading role for the U.S. in SETI.

Initiating the SETI effort

To carry on a significant United States SETI effort, public funds must be committed explicitly, with the approval of both the legislative and executive branches of the Federal Government. The evolution of an appropriate federal program lies with Congress and the President, but can only follow much preparatory work supported by one or more existing agencies.

We can recognize that successful administration of the SETI program will require leadership by an agency with: (a) a mandate to carry out scientific research and exploration, possibly requiring operations in space; (b) large scale project management experience; (c) the ability to successfully involve the U.S. and foreign scientific community in a large scale enterprise; (d) in-house expertise in the relevant fields of technology; and (e) long range goals compatible with SETI.

Since NASA clearly meets these criteria it is particularly appropriate for NASA to take the lead in the early activities of a SETI program. SETI is an exploration of the Cosmos, clearly within the intent of legislation that established NASA in 1958. SETI overlaps and is synergistic with long term NASA programs in space astronomy, exobiology, deep space communications, and planetary science. NASA is qualified technically, administratively, and practically to develop a national SETI strategy based on thoughtful interaction both with the scientific community and beyond to broader constituencies.

We therefore recommend that NASA continue its pioneering initiative in studying and planning near-term activities in support of SETI, and we urge that NASA, in cooperation with other agencies, begin the implementation of SETI.

Acta Astronautica. Vol. 6, pp. 33–46. Pergamon Press 1979. Printed in Great Britain

Strategy for the search for extraterrestrial intelligence

N. S. KARDASHEV
Institute for Space Research, Academy of Sciences, U.S.S.R.

Abstract—The problem of detection and investigation of extraterrestrial intelligence is exceptionally important for mankind from a practical standpoint, for its culture and philosophy. Its importance can even be compared to the importance of the main problems confronting our civilization at the present time, since the information obtained as a result of the discovery of intelligence in the Universe will probably furnish a practical proof of the possibility of further progress and point to ways in which our civilization could be developed over astronomically long periods of time. The use of this information could radically alter our entire concept of life and activity. Interest in solving such a superproblem is growing with each successive year while drawing on the skills of an increasing number of highly competent specialists working in different branches of natural science.

At the same time, the basic initial premises as to what we intend to conduct a search for and how to go about it are still controversial and contradictory. The purpose of this article is to attempt a formulation and critical analysis of the basic concepts and research strategies proposed by various scientists.

Evaluation of present-day state of research

(1) Several groups in the U.S.S.R. and abroad (mainly the U.S.A.) are currently conducting a search for radio signals from remote parts of the Universe in accordance with recommendations of the joint Soviet-American CETI Conference (Sagan (1973)). A program for such research has been elaborated in the USSR Academy of Sciences (CETI (1975)). The results of some of the latest research by Kardashev (1977), Dixon (1977) and Shklovskii (1976) are still discouraging. In this connection one may be left with the impression that the probability of solving this problem successfully is small (Zuckerman and Palmer, 1977). It is entirely clear to the experimenters that this opinion is unjustified. On the contrary, a new series of data indicates that the probability of solving this problem successfully is high. In addition, a very important point which is not explicitly evident in existing publications but is well known to specialists, is that all experiments that are being conducted apply at best to elaborating a search method, and even that is stretching matters a bit too far. Suffice it to say that in the last sufficiently thorough study whose objective was a search for artificial signals, about 600 of the nearest stars were observed at wavelengths near 21 cm but the time spent on studying each star was only several minutes (Shklovskii, 1976). Clearly the probability of the transmitter being turned on and the antenna being directed to Earth within the space of a few minutes on the system of planets that are studied is negligibly small.

An even more serious shortcoming of such investigations is the absence of a consistent strategy, a set of assumptions which is free of logical contradictions about what the civilizations which we are attempting to discover represent and what they can emit. In fact the most common concept about extraterrestrial intelligence is the so-called "Earth chauvinism" hypothesis which assumes that we must find beings that resemble us and who also possess modern technology on a par with ours. This viewpoint, which is totally unacceptable after a logical analysis, unfortunately still manages to survive. Apparently, at the present time, this question is the most important issue in the entire problem, and unless it is resolved, all experiments and all theoretical studies that are being conducted will not result in any advances. To obtain a more comprehensive idea about the state of the problem, in addition to performing direct experiments whose objective is the detection of artificial radio signals, data from pertinent studies conducted in astrophysics and other basic sciences which facilitate its resolution must also be analyzed. An analysis of the achievements of astrophysics over this last century implies that the systematic increase in the number of astronomical objects considered as possible shelters of life represents a leading trend in the development of the concept of a multiplicity of inhabited worlds. However at the same time, the successes achieved in biology, cybernetics, and other sciences, alter the concepts of life and civilization, which in turn necessitate a generalization of these concepts. These circumstances are decisive in the development of a search strategy for extraterrestrial intelligence.

(2) We consider it important that our first premise, namely, that there is no basis for considering the Earth and the Solar System as unique objects, be put on a firm scientific basis.

One of the most important achievements of modern astrophysics is the vast volume of data obtained from observations indicating the identity of the forms of matter over vast expanses of space extending over billions of light years and a time span of billions of years. This space includes over ten billion galaxies, and each galaxy over ten billion star systems. Because of the finite velocity with which light propagates, we see these star systems as they were thousands, millions and billions of years ago. An analysis of all this material shows that our Sun is not an isolated star by any known criteria.

Great successes have been achieved recently in substantiating the multiplicity of existing systems of planets. A theory describing the formation of systems of planets simultaneously with a sun by condensing interstellar dust and gas, is being successfully developed (Gold, 1975; Dole, 1974). This process was simulated on a computer and showed good agreement with the observed characteristics of the solar system (Dole, 1970). An important validation of the theory is the recent discovery of very narrow rings around the planet Uranus (Elliot, 1977) which form, to use a figure of speech, a miniature system of planets (the orbit sizes of the satellites and the rings obey the same laws). This confirms the prediction made by the theory, namely, that the rings around the planet are satellites that have not yet completed the condensation process. We are witnessing the process of their formation.

A large volume of data was also obtained from direct observations of the condensation process in other stellar systems. This process is studied most effectively on the basis of maser radiation in OH lines (wavelength 18 cm) and, especially, in the water vapor lines (wavelength 1.35 cm). The radio pictures of regions in which stars are formed, obtained using intercontinental radio interferometers (Batchelor, 1976; Menger and Smith, 1976), always indicate the presence of a very large number of objects undergoing condensation. The multiplicity of already formed star systems is a very common phenomenon. This also confirms the tendency toward the simultaneous formation of a large number of celestial bodies. It is important to note that in most cases, multiple star systems not only do not hinder planet formation, but rather tend to enhance this process, because resonance orbits (orbits along which the orbital period is a multiple of a stellar revolution—like the satellite orbits in the Uranus system), which are most propitious for the condensation process, can also exist in multiple systems. The origination and development of life on planets which revolve around multiple stars is also possible. For example, in the case of binary stars, when the stars form a close pair the planet may move along a stable nearly circular remote orbit. Conversely, when the stars are at a great distance from each other, the planet may revolve along a stable nearly circular orbit around either star. Again, a well known example of the second type of orbit in the Solar System is the motion of three bodies; Sun, planet, and satellite. However, for the origin of life it is important that the intensity of illumination and, therefore, also the heat conditions on the planet be sufficiently stable for both types of orbits.

We should also mention the latest results of observations of the nearest stars whose objective was the detection of planets around them. The most thorough and uniform observation of changes in the positions of Barnard's star made during 1950–74, confirm the presence of at least two satellites with masses whose order of magnitude is 1 and 0.4 of Jupiter's mass (Van de Kamp, 1975). Thus, there is no basis for doubting the great abundance of systems of planets with suitable conditions for life.

The detection in astronomical objects of complex organic molecules which are considered to be indispensible for the origin of life is also a major discovery. Emission and absorption lines indicating the nearly ubiquitous presence of organic compounds in interstellar clouds have been discovered, mainly by radio methods. New and often more complex molecular compounds are being discovered every year. These compounds can reach the surface of planets while the latter are undergoing condensation, so that there is no reason for doubting the availability of ample starting material for the origin of life.

We now pass on to the problem of the origin of life. It is a well known fact that the mechanism responsible for the origin of life is not yet understood. However, available data indicate that this mechanism is a physicochemical processes which obeys strict laws describing the behavior of complex organic molecules; and it ensured the very fast appearance of a universal genetic code in the early stages of the Earth's evolution (Tze-Fei Wong, 1976). Therefore, attempts aiming to prove a small probability for the event that a living cell will be formed as a result of random collisions of individual atoms and molecules must be rejected.

An explanation of the original synthesis of living matter is an exceptionally difficult problem since, apparently, all traces of this process have been erased on Earth by subsequent biological evolution. The general consensus is that we are living at the threshold of a revolution in biology. The first steps in this process were closely related to understanding the heredity code mechanism revealed during a study of cell structure on the molecular level. However, at the same time, it is clear that the biological cell is a relatively recent development and that it is probably easier to learn to control an existing cell than to understand what it was like billions of years ago, and its origin. In precisely the same way, the evolution of life, the development of civilization, and its transformation into a technologically highly developed society, also appears to obey laws although, in this field, we are still very far from a rigorous scientific theory describing all the stages of its development.

(3) In concluding this section we present some of the latest estimates on the expected number of civiliations in our galaxy. During a discussion of the "Cyclops" project (1972), the following magnitudes were assumed: rate of star formation in galaxy: 20 stars per year (among these the proportion of stars with planets: 50%); for each system of planets, the number of planets suitable as a habitat: 1 planet; among these, the proportion of planets on which life develops: 20%; proportion of planets with some form of life on which intelligent life appears: 100%; proportion of planets with intelligent life and a developing technological civilization desiring to establish contact with other civilizations: 50%. These figures give the expected number of civilizations existing simultaneously in the galaxy, which turns out to be numerically equal to the length (measured in years) of the average existence of each civilization in the phase during which it is capable of establishing contact with other civilizations. This quantity is unknown. It can be a magnitude which is comparable to the age of the galaxy (i.e. billions of years) and billions of civilizations existing simultaneously in our star system. However, if one agrees with predictions of a quick destruction of our civilization (within decades—according to well known estimates by the Club of Rome), the corresponding number of simultaneously existing civilizations in the galaxy can be several tens. It is possible that the values of the proportions presented above are somewhat smaller. For example one of the most recent studies (Hohlfeld and Terzian, 1977) gives a value of their product which is fifty times smaller. Therefore, if we assume a brief span for the duration of life, we can present an argument in favor of the proof of the uniqueness of life on Earth (Zuckerman and Palmer, 1977). However this concept is totally unacceptable. There is no objective basis for the assumption that some fatalistic law is operating in the Universe to destroy every civilization several decades after it has entered the stage in which it can communicate, and that this law already has begun to manifest itself on Earth. Undoubtedly the possibility exists that civilization on Earth will be destroyed as a result of war devastation or depletion of natural resources. This is bound to occur unless combatting these problems becomes a basic part of mankind's activity in the nearest future. Taking into consideration the seriousness of the current situation, we cannot deny the possibility that a constructive solution to the problem of preserving civilization on Earth will not be found. It is precisely this problem and the estimates of the duration of technological, cultural, and scientific development of society associated with it, which are decisive factors in the

elaboration of a search strategy for civilizations. It was already noted above that further research cannot be conducted without discussion and adoption of a logical, noncontradictory model for the development of civilizations.

Whom can we find?

We present here some additional concepts that were discussed earlier in Dole (1974), and attempt to formalize the concept of a civilization, keeping in mind the fact that our knowledge in this field is still very meager. At the same time, we do not doubt the possibility of and usefulness of a generalizing concept pertaining to any sphere of activity in our civilization. Starting with the main factor, the uniqueness of all basic laws of nature in the entire space observed throughout the evolution of our civilization, as reliably established by astrophysics, we are justified in assuming that the natural laws known to us are also known and used by other civilizations. It is important to note that our civilization, capable of establishing contact with other civilizations, is still very young, and that its age represents an order of magnitude which is very small, possibly zero. Taking into account the fact that the Solar System is a second generation object, that its age is about 5 billion years, and that the age of the oldest objects in the Universe can be about 20 billion years, it becomes clear that the age of other civilizations (in particular the time period throughout which they have been communicating) can be enormously greater than ours. Therefore they may know much more than we do; at any rate, their knowledge must comprise all our knowledge. Moreover, again on the basis of the uniformity of the forms of objects and laws governing their evolution in the entire observed space of the Universe, it is highly probable that our knowledge represents a necessary stage in the initial development of any civilization. This common factor allows us to attempt a definition (at least in very rough approximation) of a civilization. The functional definition of the concept of life on Earth given by A. A. Lyapunov (1963), namely "a highly stable state of matter, utilizing information encoded by the states of individual molecules for the purpose of developing reactions aimed at self-preservation," can serve as a basis for this definition.

This definition reflects sufficiently accurately the essentials of the concept; however, in our opinion, it has a serious drawback—it says nothing about the laws by virtue of which the origin, development and increasing complexity of living forms occurs. Apparently, the life history of each individual can be considered as a stochastic process which is determined by particular statistical laws. Both are determined by interaction with the environment and by the natural state of the living organism at each instant. The evolution of species also represents a stochastic law in the development of individuals. The common outcome of evolution is a process in which information and its possible uses are accumulated. Therefore it appears that a basic statistical law which applies to living organisms is their striving to acquire and utilize a maximum amount of information about their surroundings and themselves. It appears very likely that life originates as a result of a possible synthesis of special kinds of molecules capable of storing and utilizing in the beginning the simplest information about the environment and their own structure. The latter in turn is used for self-preservation, reproduction, and the acquisition of a still larger volume of information which is of particular importance to us. This law manifests itself as a result of the action of inherent physical and chemical forces in these molecules and, also, depends on environmental conditions. In conformity with biological terminology, the law for lower forms of life is realized through mutations and natural selection. One would think that the same law is the sole stimulus for the development of higher forms of civilizations.

An essential difference distinguishing higher forms of life is their ability to abstractly analyze the collected information. In the course of the development of various forms of life, systems of living organisms reflecting this law also begin to play an important part in evolution. Using the definition of life given above, the following functional definition of a more advanced civilization can be proposed: "a highly stable state of matter capable of collecting, abstractly analyzing, and using information for the purpose of acquiring qualitatively new information about its surroundings and itself, for self-improvement from the standpoint of possible acquisition of new information, and for the development of self-preserving reactions, which is uniquely distinguished by the amount of collected information, the program for its functioning and by its production for realizing these functions." Concrete specification of a method for encoding the information seems inappropriate in the given case since it would have to include books, image-recording equipment, the memory of an electronic computer, and possibly, artificial forms of life which may be created in the future. By information about the surroundings and itself is meant knowledge about lifeless and living nature (including civilization).

Figure 1 clarifies the definition given above. It shows graphically the interaction of the civilization with the Universe by which it is surrounded. To realize the goals that were set, the civilization utilizes production and its total accumulated knowledge consisting of the available information in the natural and social sciences and humanities. It also uses a program which analyzes this information and issues instructions which control production. The definition of a civilization presupposes that work is not a goal in itself, but a means for realizing the functions enumerated in the definition.

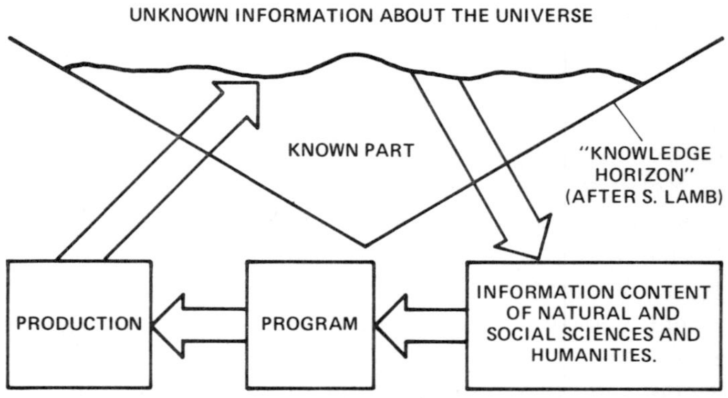

Fig. 1. Interaction of a civilization with the surrounding Universe.

The next important problem is the concept of the structure and hierarchy of levels of information about nature and society. Frequently the concept of an information tree is introduced, in which only isolated branches are known to us at a given stage of development. This is precisely the reason why the definition of a civilization emphasizes that continuous acquisition of qualitatively new information, or information of a higher order generalizing known laws, or the discovery of new hitherto unknown branches of the information tree, is indispensible for further development. In this connection, there arises the question of the relationship between the length of development of civilizations and the finite or infinite number of fundamental natural laws that are still unknown. If we are approaching a state in which all important laws are known, only quantitative development is possible in the future. In this we can also see the possible destruction of mankind due to boredom (degeneration, loss of interest). However, this problem does not exist, at least for the time being. We present several examples of such problems that have not been resolved so far.

In natural science:

- unified theory of gravitation and relativistic quantum mechanics
- theory of elementary particles
- a theory explaining the numerical value of fundamental physical constants—the velocity of light, Planck's constant, the gravitational constant, charge and mass of an electron and of a proton (or a theory explaining the meaning of their dimensionless combinations)
- the problem of the finiteness or non-finiteness of the diversity of fundamental laws for the microcosm
- A theory explaining what existed at the beginning of and before the observed expansion of the Universe
- a theory explaining the numerical values of the rate of expansion and acceleration of the Universe, its mean density and entropy
- a theory showing what takes place inside a black hole for an observer drawn into it
- is outer space simply connected? If not, do other universes exist besides the observed micro and macrocosm?
- is the number of basic laws governing the macrocosm finite?
- a theory of the origin of life on Earth and forms of life's origin in other parts of the Universe
- a theory of brain activity in man and animals.

In the social sciences:

- a theory of optimal economic and social organization with the aim of ensuring development, avoiding war and protecting the environment
- what should be given first priority—colonization of outer space around Earth by Man, or ordering life on Earth?
- how should outer space be colonized—separately by socialist countries, separately by capitalist countries, or jointly?

- when artificial intelligence appears, what will society with artificial intelligence be like?
- how will the evolution of civilization depend on the discovery of new fundamental laws in natural science?

Thus, a set of very important problems exist whose solution cannot be predicted. Several conclusions follow from this.

First conclusion. Civilization is confronted with a vast, probably unlimited field of activity connected with the solution of problems which are similar to the examples presented above. Therefore, the duration of development and the phase in which civilizations communicate is very long, possibly unlimited.

Second conclusion. Our contemporary stage of development represents only an insignificant part of this phase, so that it is highly improbable that we shall encounter "intelligent fellow beings" at the same stage of development as ours.

Third conclusion. Highly developed civilizations understand and apply on a wide scale, laws which are still unknown to us. In my opinion, this last circumstance must also be taken into consideration in an elaboration of a search strategy for extraterrestrial intelligence.

Next, it appears expedient to analyze several hypothetical models of the evolution of civilizations.

(1) Surprise is expressed by Zuckerman and Palmer (1977) as to why the "intelligence shock wave" of one of these supercivilizations did not encompass the entire Universe. Taking into consideration the concepts presented above, the following objection can be raised to this possibility: expansion of occupied space is by no means a necessary part of the activity of a civilization. First, if everything is the same in the Universe, why expand? Second, a larger system is inconvenient from the standpoint of the rate at which information is exchanged between its parts. On the contrary, it is more advantageous if two civilizations located at a great distance from each other combine. This increases the amount of information possessed by each civilization. However, then, the number of civilizations and the space occupied by them will be even smaller.

(2) Instead of a model of unbounded space expansion, the following possible paths of activity of civilization can be outlined with the prospect of obtaining information about new fundamental laws in the future:

- investigation of microcosm
- goal oriented space flights to the most interesting objects in our macrocosm (for example, should we fly to the center of the galaxy in the future?)
- study of the possibility of making a change to other types of measurements (for example through a charged black hole (Kardashev, 1975)).

We can assume that civilizations use large quantities of energy for such activitities. It is also natural to expect a sufficiently efficient use of this energy. Then, on the basis of thermodynamic concepts, the energy used up is mainly converted into radiation whose intensity (measured by a bolometer) is approximately equal to the intensity of the background radiation surrounding the source. The spectral distribution of this intensity will be roughly the same as that of a black body. This provides one possibility for the search for civilizations. The second natural assumption that should be verified by observations is the large volume of solid material used by a civilization for astroengineering purposes. Finally, the last assumption is that signals carry information in the form of electromagnetic radiation.

The examples presented above of course do not exhaust all possible directions of civilized activity. However, evidently, all of them are directly related to the problem of elaborating a strategy for the search for civilizations, to which we can now pass on immediately.

How should the search be conducted, and how should it not be conducted?

Scientists are currently discussing two concepts and two corresponding but diametrically opposed strategies for the search for extraterrestrial civilizations.

Earth chauvinism concept

Every civilization stabilizes or perishes in a developmental stage which is close to ours (see the conclusions and recommendations of the Club of Rome; tight restrictions on population size and its activity). *Strategy.* Search for civilizations which are similar to our civilization; i.e. civilizations that form on part of a system of planets and use communications techniques which are similar to ours. Search for monochromatic signals from all the nearest stars.

Evolutionary concept

The possibility of a much higher level of civilizations than our contemporary stage of development is assumed. *Strategy.* Search for supercivilizations possessing more powerful transmitters and/or with highly developed astroengineering activity.

Investigations are conducted in two directions. First, the nature of the most powerful (and frequently most remote) sources of radiation in the Universe is studied. A transition to an investigation of less powerful objects in the Universe is expedient only after the physics of the most powerful sources have been definitely established. Second, a search for new powerful sources of radiation is conducted, especially in regions of the electromagnetic spectrum that have not been the objective of extensive study so far.

It is expedient to search for monochromatic or pulsed communication signals from the center of the galaxy, galactic and quasar nuclei and other classes of astronomical objects (if there are grounds for linking them to the development of civilizations); to seek traces of astroengineering activity in images of these objects obtained with superhigh resolution; and to study in detail peculiar sources already detected (for example, very large solid structures can be detected, some of which may be moving at velocities approaching the velocity of light).

In our opinion, observations and the comparison and analysis of all scientific data that have been made, point unmistakenly to the logical contradiction and lack of promise of the first approach and to the necessity of concentrating efforts on the second possibility. Chief attention must be focused on the search for new objects and the investigation of galactic nuclei and quasars. Here two hypotheses are possible. One hypothesis presupposes that the galactic nuclei themselves, and quasars, are related to the activity of civilizations. The second hypothesis is that civilizations utilize the enormous radiation flow emitted by galactic nuclei and quasars in a way similar to the way we utilize solar energy. According to the latest data, the radiation intensity of the Galactic nucleus is 10^6 times, and that of a quasar 10^{12} times, greater than the luminosity of the Sun. Apparently one can always imagine large astroengineering structures located, for the sake of safety, at a particular distance from these superpowerful sources of energy. Since the densities of matter in quasars and galactic nuclei are not high, it is probably also possible that the structures are located in them. If the radiation is used very efficiently, these structures are sources of nearly black body radiation—a fact already mentioned. The temperature will be in the range of 2.7°K (equilibrium with relict background) to 300°K (room temperature), if this temperature is required for normal operation of these structures, or if the intensity of the surrounding background radiation exceeds that of the relict radiation.

The development of civilization on Earth at present rates, even when extrapolated to time periods which are much shorter than those in the Universe, corresponds to the energy levels of quasars and galactic nuclei. It is important to note that in accordance with the concepts of astrophysics, these sources are objects that evolved in the first generation during the evolution of the Universe (simultaneously with the formation of the first stars in galaxies).

The second first generation components are subdwarfs representing about 20% of all stars in our galaxy (20 billion stars). The main feature of subdwarfs is their spherical distribution and concentration near the center of the galaxy. The orbits of these oldest galactic objects are nearly radial (they pass very close to the galactic nucleus), and the age of subdwarfs (and consequently the possible age of civilizations engendered on them) is 10–15 billion years longer than that of the Solar System. Therefore a complete comprehension of the phenomena occurring in the spherical component of the stellar population and in the nucleus is simultaneously the object of investigations related to the search for civilizations in our Galaxy and the most important problem of astrophysics.

The optimal range of wavelength for the search for artificial signals from the nucleus of our Galaxy, the nuclei of other galaxies and quasars, differs substantially from the optimal band for communication with the nearest stars. These objectives, whose nature is still unknown, are powerful emitters of synchrontron radiation throughout the entire band of radio frequencies. In addition, the scattering of radio waves by the plasma clouds surrounding these objects has a strong effect on the emissions. Taking into account these circumstances, it is probable that the range of maximum intensity of the relic background radiation, not the 21 cm line, is optimal. It is of interest to note that if a civilization located on a distant extragalactic object emits signals in this range, observers in our Galaxy will also record signals in the range of maximum relic background radiation intensity which they observe, regardless of the change in the spectrum of the background radiation which occurred during propagation of the signal.

For us the range of maximum intensity of the relic radiation background corresponds to a wavelength of 1.7 mm. We can make an attempt in this range to detect special signals carrying information and thermal radiation from large astroengineering structures. In this connection, we need an understanding of what the point radio source located exactly at the centre of the Galaxy represents (Davis, Walsh, and Booth, 1976; Kellerman, 1977). It is smaller than the dimensions of the Solar System and emits shortwave radiation. And a clarification of the several sources of infrared radiation around the center whose temperature is close to room temperature (Ricke and Low, 1973; Becklin and Neugebauer, 1975), is of exceptional interest. We need to study the region of the center (Eyles, Skinner, Willmore, and Rosenberg, 1975) which was resolved at the beginning of 1975, since we can reckon with a technology and with safety measures that exceed immeasurably our present day capabilities.

But are we not in danger of losing our common sense when we attribute such content to actually observed astrophysical phenomena?

At the present stage of development, it is probably most expedient to keep in mind simultaneously two possibilities for objects whose nature is unknown— a "natural" explanation of objects formed as a result of the evolution of a lifeless Universe, and objects which can be labeled as a "cosmic wonder", that could have originated as a result of sustained evolution of intelligent life in the Universe. Both concepts can be used to set up and carry out experiments which will clarify what assumptions come closer to the truth. Therefore, the "presumption of natural origin" for every astronomical object appears to be totally unacceptable. This presumption is a violation of creative effort; and every scientist has the right to work within the framework of his own systems of ideas and intuition.

An additional important circumstance must be kept in mind while discussing the possible existence of supercivilizations and a program for their search. It is possible that their activity is essentially related to tasks which we do not even suspect. It was already mentioned above that, for example, a very important and completely unresolved problem is the question of the possibility of our space being multiply-connected. Is it not possible that after some period of development every civilization learns "everything about its own universe" (taking into consideration that all its properties are the same in all directions and over time)? Then the only way out of the impasse is to establish contact with other supercivilizations and, if possible, departure to another space. Perhaps this is the reason why no traces of activity of other civilizations are seen on Earth and in the immediate vicinity of the Solar System. All these are of course only assumptions showing the diversity of situations encountered in an attempt to solve the search problem.

Finally we present basic estimates for the possibility of detecting extraterrestrial civilizations by means of modern engineering equipment. These calculations are very reliable, since they are based on known conditions under which electromagnetic signals propagate in outer space. According to a study (Buyakas, Gvamichava *et al.*, 1977), a complex of space radiotelescopes is being consider with a mirror diameter up to 10 km. The calculations show that the range at which civilizations can be detected on the basis of free emission from astroengineering structures, the absorption or reflection of background radiowaves, or on the basis of information carrying signals, increases to a limiting value which is comparable to the radius of curvature of the universe (depending on the diameter D of the antenna). These conclusions are illustrated in Figure 2. Estimates of the detectability at millimeter wavelengths of astroengineering structures ("Dyson's sphere" at temperatures 3°K and 300°K) or communication signals are shown, assuming a very modest energy use by the supercivilization (on the order of one solar luminosity). The possibilities of detecting transmitters which are similar to those existing on Earth are also indicated in the figure ("Earth chauvinism concept"). The detection range for the evolutionary concept, given the same research equipment, is greater by many orders of magnitude.

To end this discussion, we present the main conclusions. Astrophysical research, data from biology and cybernetics, and from other sciences, point to a high probability for the detectability of extraterrestrial civilizations. Currently, what is needed is a fundamental review of our preliminary notions about the possible nature of these civilizations and of the particular method which must be used in the search for them. In my opinion, the only useful concept is the assumption that supercivilizations exist (in particular, also, that our civilization may eventually become a supercivilization). The chief method to be used for their detection consists of carrying out a series of experiments. First, a search must be carried out by investigating black body radiation sources with peak intensity in the millimeter and infrared wavelength ranges. The object here is the detection of very large structures. It is possible that these structures can also be detected on the basis of their screening action, or on the basis of the cosmic background radiation reflected by them in various spectral ranges. Here, image

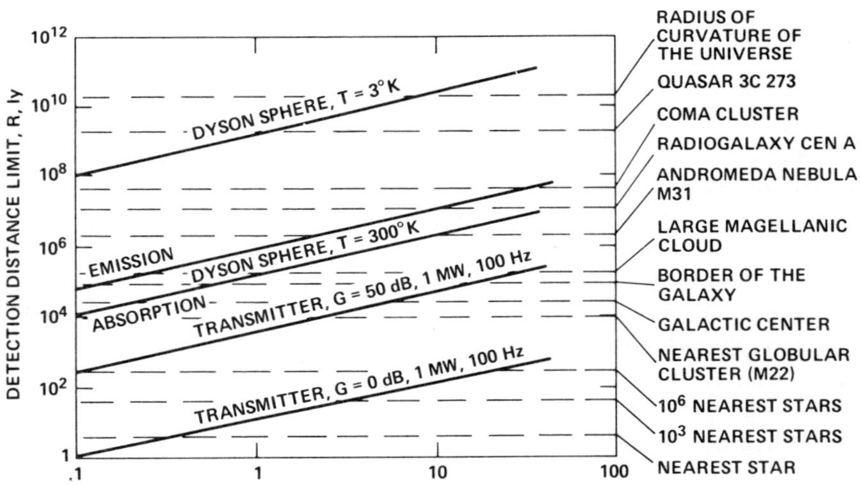

Fig. 2. Estimates of the detectability at millimeter wavelengths of astroengineering structures and communication signals; range versus the diameter D of a filled aperture antenna.

synthesis of the sources that are being studied, using interferometers with a very long baseline, or detection of very compact objects by the scintillation method (i.e. the modulation of radiation caused by the passage of a signal through the inhomogeneous cosmic plasma), appear to be very promising. To obtain ultra-high resolving power in the interferometric mode and during observations by the scintillation method, the radiotelescopes must be spaced at distances which are greater than the Earth's diameter. Second, a search must be conducted for communication signals from suspect objects. The optimal wavelength range for such signals must be calculated taking into account the background radiation in the vicinities of the source and the conditions under which signals propagate in the cosmic plasma. The millimeter wavelength range, which includes the maximum intensity of the relict background radiation, appear to be most promising. It is assumed that these experiments will either reveal a connection between known astronomical objects and the activity of civilizations, or that they will make it possible to discover a completely new class of such sources. Estimates of the possibility of a discovery indicate that by performing carefully prepared experiments based on an evolutionary concept which is free of logical contradictions and in which the largest modest radio telescopes are utilized, a project for the search for extraterrestrial civilizations can produce favorable results within the next decade. Also, in this way, the immense volume of information stored in the Universe over billions of years will be accessible to mankind.

References

Batchelor R. *et al.* (1976) *Pis'ma V.A.Zh.* **2**, 467; Walter R., Johnson K., Burke B. and Spender J. *Astrophys. J.* **211**, L135.

Becklin E. and Neugehauer G. (1975) *Astrophys. Jour.* **200**, L71.

Buyakas V. I., Gvamichava A. S. *et al.* (1977) Neogranichenno narashchivayemyy kosmicheskii radio teleskop (Space Radio Telescope that Can Be Enlarged Without Limit), Report read at the International Austronautical Federation Congress in Prague. *Acta Astronautica* **6**, 175–201.

CETI Program, The, (1975) Sov. Astron., Vol. 18, No. 5, Mar.–Apr.: *Astron. Zhur.* **51**(5), 1125 (1974).

Davies R., Walsh D. and Booth P. (1976) *MNRAS* **177**, 319.

Dixon R. and Cole E. (1977) *Icarus* **30**, 267.

Dole S. (1974) *Planety dyla lyudey* (Planets for People), Nauka Press, Moscow.

Dole S. (1970) *Icarus* **13**, 494.

Elliot J. and Dunham E. (1977) *Sky and Telescope* **53**, 412.

Eyles C. J., Skinner G. K., Willmore A. P. and Rosenberg F. D. (1975) *Nature* **257**, 291.

Gold T. (1975) *Problemy CETI* (CETI Problems), pp. 15–26, Mir Press.

Hohfeld R. and Terzian Y. (1977) *Icarus.*

Kardashev N. S. (1969) *Vnezemnyye tsivilizatsii* (Extraterrestrial Intelligence) pp. 43–52, Nauka Press, Moscow.

Kardashev N. S. (1975) *Problemy CETI* (CETI Problems), pp. 166–172, Mir Press.

Kardashev N. S. *et al.* (1977) *Astron. Zhur.* **54**(1), 3.

Kellermann K. *et al.* (1977) *Astrophys. Jour.* **214**, L61.

Lyapunov A. A. (1963) *Problemy kibernetiky*, Issue 10, p. 179.

Menger P. and Smith L. (1976) *IAU Symposium No.* 75.

Problemy CETI (CETI Problems) (1975) Mir Press, p. 319.

Project Cyclops (1972) NASA Ames Research Center, CR114445.

Ricke G. and Low F. (1973) *Astrophys. Jour.* **184**, 415.

Sagan C. (1973) *Communications with Extraterrestrial Intelligence*, p. 352. M.I.T. Press.
Shklovskii I. S. (1975) *Problemy CETI* (CETI Problems), p. 133. Mir Press.
Shklovskii I. S. (1976) *Voprosy filosofii* **9**, 80.
Tze-Fei Wong J. (1976) *Proc. Nat. Acad. Sci. U.S.A.* **73**(7), 2236.
Van de Kamp (1975) *Astron. J.* **80**, 658: (1975) *An. Rev. Astron. and Astroph.* **13**, 295.
Zuckerman B. and Palmer P. (1977) *Icarus*.

Acta Astronautica. Vol. 6, pp. 47–57. Pergamon Press 1979. Printed in Great Britain

A review of the theory of interstellar communication

J. BILLINGHAM,† B. M. OLIVER‡ AND J. H. WOLFE§

Ames Research Center, NASA, Moffett Field, CA 94035, U.S.A.

Abstract—Sixteen years have elapsed since Cocconi and Morrison suggested that attempts might be made to detect communications from extraterrestrial civilizations. In this time there have been three major types of activity. First, the general theory of the estimation of the number of technological civilizations has been developed and expanded. Second, studies have been made of the designs of systems of sufficient sensitivity to give reasonable chances of intercepting signals. Third, attempts have been made, on a small scale, to listen specifically for intelligent signals. Two additional events should be mentioned: the Pioneer 10 spacecraft, the first man-made object to escape the solar system, is carrying a plaque depicting some basic data about Earth and human life; and a signal with similar information was recently sent from the Arecibo telescope to the Messier 13 galaxy. In view of this activity, it is reasonable to ask whether there are now grounds for suggesting that a serious attempt be made to design a system to search for extraterrestrial intelligent life. Such grounds may indeed now exist.

The number of coexisting technological civilizations is related, in the formulation proposed by Drake, to the rate of star formation in the galaxy, the fraction of stars presently having planetary systems, the suitability of the stellar system, the number of planets in the life zone of each star, the probability of the emergence of life on suitable planets, the fraction of biological systems becoming intelligent, the number of intelligent species attempting communication, and the average lifetime of a communicative civilization. The last factor appears to have the greatest uncertainty.

Since the original formula was proposed, there have been a number of advances. It has been proposed by one of us (Oliver) that situations will have arisen in galactic history where a civilization is sufficiently close to other planets on which life has developed that discovery of that life is highly probable. These situations are of four types: first, civilizations could arise independently on the planets of single stars that are separated by only a few light years. Even if the lifetimes of the civilizations are measured only in thousands of years, there will still have been a significant number of instances of coexisting communicative civilizations which are likely to have discovered each other. Second, civilizations could have developed simultaneously on planets around binary stars, with an even greater likelihood of contact. Third, a civilization on a planet of one star of a binary system might have traveled to planets around the second star and discovered non-intelligent life. Fourth, a civilization in a single solar system might have discovered non-intelligent life on a planet in the same system. This would be equivalent to our own discovery of primitive or extinct life on Mars.

The probability that one or more of these discoveries has taken place is high. Such a discovery would provide a strong stimulus to the civilization in question to attempt to contact more distant civilizations, with the evolution of a network of communicating cultures across the galaxy. It is possible that the rapid increases in knowledge facilitated by such communication would accelerate the evolution of the participants to the point where they are able to achieve stability and longevity.

---

†Chief, Interstellar Communication Study Group.

‡Vice-President for Research and Development, Hewlett–Packard Corporation, Palo Alto, CA 94304 U.S.A.

§Deputy Chief, Interstellar Communication Study Group.

In contrast, those civilizations which do not attempt communication may retrogress. The achievement of interstellar communication, with subsequent membership in the galactic community, could represent a decisive step in the continuation of biological evolution.

Attempts have been made with existing radiotelescopes, over the last 15 years, to detect coherent signals characteristic of intelligent life. Project Ozma, in 1960, was the first attempt. Subsequent experiments are described. None has been successful to date, which is perhaps not surprising in view of the low sensitivity of the receiving system and the limited frequency range employed. Most of the activity has been in the microwave region of the spectrum. Other approaches, such as laser communication, interstellar probes, or the detection of infrared emissions from stellar system engineering projects, have been suggested and will be reviewed.

The first attempt to study in detail the design of a system for contacting extraterrestrial intelligent life was made in the summer of 1971 by a team of scientists at the NASA Ames Research Center, under the direction of two of us (Oliver and Billingham). The concept, known as Project Cyclops, is a large array of phased radiotelescopes on the Earth, listening in the microwave region of the spectrum for coherent signals from star systems within 1000 light years. The microwave region was chosen because galatic and quantum noise are at a minimum, and because the spectral lines of hydrogen and hydroxyl (1420 and 1680 MHz) fall in the center of the region. The data-processing system examines simultaneously all 0.1-Hz channels within this band and can detect signals that are 90 db below the integrated noise level in the band. The phased array would be pointed at each likely star for a period of 20 minutes, and then be switched to the next star. A few million candidate stars are available within 1000 light years, and all could be examined within a few decades by the Cyclops system.

In conclusion, it is suggested that the discovery of extraterrestrial intelligent life could be of such profound importance to mankind that further studies should be carried out. The technology is available, and the costs seem reasonable in comparison with other major endeavors in science and exploration.

Occasional speculation about the existence of life, and even of intelligent life, on worlds other than our own, goes back over the centuries. In the last two decades, there has been a rapid advance in the theory of interstellar communication, culminating most recently in the first studies of systems designed specifically for the detection of extraterrestrial intelligent life. A comprehensive bibliography of interstellar communication was recently published by Mallove and Forward (1974, 1975).

The first half of this century saw the development of the concept of chemical evolution, whereby organic compounds similar to those which make up living systems today were first postulated (Haldane, 1954; Oparin, 1961) and then demonstrated (Miller, 1955) to be synthesized in chemical environments similar to those of primitive Earth. By now, amino acids, sugars, purines, pyrimidines and fatty acids have been demonstrated in such experiments (Miller and Orgel, 1974). In 1960, Lederberg coined the term *exobiology* to describe the study of life outside the Earth. The theory envisages a gradual assembly of organic compounds to form the first self-replicating systems and eventually the earliest living cells. In the United States, two Project Viking spacecraft have now been launched to search directly for evidence of microbiological life on Mars (Corliss, 1974).

General scientific discussion of the fact that life might be widespread in the Universe, and that such life may become intelligent, is of more recent origin. In 1959, Cocconi and Morrison published their paper "Searching for Interstellar

Communication" in Nature. They proposed the examination of the region of the neutral line of hydrogen (1420 MHz) as a means by which civilizations might communicate with each other over interstellar distances. In 1963, a collection of papers by Bracewell, Dyson, Morrison, Oliver, Sagan, Shklovskii *et al.*, was published by Cameron (1963) under the title *Interstellar Communication.* In 1966, Shklovskii and Sagan published their comprehensive analysis of *Intelligent Life in the Universe.* In 1971, there appeared a translation of the Soviet collection of papers under the heading *Extraterrestrial Civilizations: Problems of Interstellar Communication,* edited by S. A. Kaplan. *Project Cyclops: A Design Study of a System for the Detection of Extraterrestrial Intelligent Life* was published in 1972 (1973, revised edition) by Oliver and Billingham. In 1973, the proceedings of a joint meeting between the United States, Soviet, and Armenian Academies of Science were published under the title *CETI* (*Communication with Extraterrestrial Intelligence*) by Carl Sagan. In 1974, a series of papers originally presented at the NASA Ames Research Center in 1970 were published by Ponnamperuma and Cameron under the title *Interstellar Communication: Scientific Perspectives.* A recent summary of the state of knowledge in the field of interstellar communication was published by Sagan and Drake (1975).

The references listed above are a few of the key publications that have appeared in recent years. They deal with various aspects of the theory of interstellar communication and with concepts for methods by which the detection of and communication with extraterrestrial civilizations might be attempted.

Our present understanding of the theory is best illustrated by reference to the equation for the prevalence of communicative civilizations first proposed by Drake:

$$N = R_* f_g f_p n_e f_l f_i f_c L \tag{1}$$

where R_* = rate of star formation; f_g = fraction of stars that are "good" suns; f_p = fraction of stars that have planetary systems; n_e = average number of "good" earths per system; f_l = fraction of good earths upon which life begins; f_i = fraction of life starts that evolve intelligence; f_c = fraction of intelligent species that develop interstellar communication; L = longevity of communicative phase.

R_* is reasonably well known. There are as many as 2×10^{11} stars in the Galaxy and its age is about 10^{10} years. Thus the average rate of star formation has been 20 per year. The rate was greater in the beginning, is less now, and was about equal to the average when the Galaxy was half its present age. Thus, $R_* \approx 20$.

We now make the chauvinistic but pessimistic assumptions that life needs Earth-like conditions to begin and that the time required to evolve intelligence on Earth is typical. If life can arise and evolve under conditions widely different from those of Earth, this only increases the probability of its occurrence.

Under these assumptions, good suns are limited to Population *I*, main sequence, *F*, *G* and *K* single stars. Population *I* stars are good because they had enough heavy elements in their nebulae to have allowed rocky planets with iron cores to form. The main sequence is good because this is the long stable period of a star's life during which it steadily burns hydrogen to helium to produce sunshine. When a star enters the giant phase, it almost certainly destroys all life on its planets. Stars with spectral range *F*, *G* and *K* are good because larger stars have too short a main sequence lifetime (<2 billion years) while smaller stars have such low luminosities that a planet, to be warm enough, must be so close that tidal coupling probably stops its rotation. Single stars are good because stable orbits exist only close to the components of widely spaced binaries or far from closely spaced binaries. Further, it is not clear that binaries will develop planets. These restrictions establish $f_g \approx 0.05$.

The fraction of stars with planetary systems is not known. Modern theory strongly suggests that they should be the rule around single stars. We have evidence for unseen companions around most stars small enough and near enough for perturbations to be detectable. We can assume on this basis that $f_p \approx 1/2$. We note in passing that 40 years ago most astronomers would have assigned f_p a value as low as 10^{-9}.

If the spacing of planets in the solar system is typical and if the Bode-Titius law holds generally, there will typically be a planet at the right distance to have a "temperate" climate for part or most of its surface. However, the size of the planet is also important if the volcanic origin of the atmospheres of the terrestrial planets is correct. Too large a planet might produce too much atmosphere, too small a planet, none. We will assume that $0.02 < n_e < 2$.

Exobiologists now believe that, given an Earth-like planet with its primitive atmosphere, life is almost certain to begin. That is, current thinking places $f_l \approx 1$. This, too, is in great contrast with the situation 40 years ago when nothing was known about DNA and the natural formation of its chemical precursors.

The fraction of living systems that evolve intelligence depends upon time and the competition for survival imposed by the environment and by rival species. We have allowed for 5 billion years in our elimination of stars more massive than type F5. Life on Earth displays the ability to adapt to the most extreme conditions and to feed on the most diverse energy sources, including other life forms. The threat posed by other life and the advantages of sensory organs and intelligence for survival are probably ubiquitous. We feel we can safely assign f_i the range $0.01 < f_i < 1$.

To become capable of interstellar communication, a species must develop a technology (such as radio astronomy) and the determination to make the attempt. Our technology grew out of attempts to develop weapons, to grow food, to produce clothing and shelter, in general to increase our safety and well-being by devising beneficial modifications to our environment. Rudimentary technology and curiosity led to science, and science led to more technology in an iterative sequence. The final step. attempting to achieve contact, is a policy decision for a civilization. We still do not know if we will make a major effort at applying our own technology to the search for other life. We are about to contribute a statistic to f_c. For the present, let us also take $0.1 < f_c < 1$.

The product $R_* f_g$ in (1) is on the order of unity. The remaining selectivity factors we have discussed can be combined into an overall factor:

$$F = f_p n_e f_l f_i f_c \tag{2}$$

which, under the assumptions for the ranges of the individual factors given above, would lie between the extremes 10^{-3} and 1. However, each of the individual selectivity factors should really be assigned an *a priori* probability density function. If, in our present state of ignorance we assume these functions are uniformly distributed on a logarithmic scale between the limits given for each factor, then the probability density function for F will be bell-shaped, with its peak at $F = 10^{-3/2}$. This would be both the most probable and mean value of F. For the assumed distribution we have

X	Probability that $F < X$
10^{-3}	0
$10^{-5/2}$	0.021
10^{-2}	0.166
$10^{-3/2}$	0.5
10^{-1}	0.833
$10^{-1/2}$	0.979
1	1

The point is that when we multiply a number of factors, to each of which, out of ignorance, we can only assign a probable range, the distribution of their product is apt to center at the product of the midpoints of the ranges and be very low at either extreme. In the above case, the probability is only 16–2/3% that $F < 0.01$ and only 2% that $F < 0.003$. Thus it appears likely that

$$N \approx FL \tag{3}$$

where $0.01 < F < 0.1$.

If we combine (1) and (2) and divide both sides by the number of stars in the Galaxy, N_*, and by the fraction f_g that are good suns:

$$\frac{N}{N_* f_g} = \frac{R_*}{N_*} FL = \frac{R_* T}{N_*} F \frac{L}{T} \tag{4}$$

Now $N_* f_g$ is the number of good suns, so the left side is the probability p that a good sun, chosen at random, will be communicative. If T is the age of the Galaxy, then $R_* T = N_*$ so

$$p \approx F \frac{L}{T} \tag{5}$$

Equations (3) and (5) enable us to compute how far into space we must search for a reasonable chance of success and how many good suns must be searched.

Our uncertainty as to the longevity L of the communicative phase is very great. Some argue pessimistically that the lifetime of a technologically advanced civilization may not exceed 10^2 or 10^3 years in its entirety, let alone the communicative phase. At the other extreme, the proponents of interstellar communication argue that the communication itself may confer longevity by teaching stable social patterns and societal forms, by motivating civilizations to long-term goals and causing them to radiate beacons to attract emerging civilizations to the Galactic community. For these reasons, L could be as great or greater than 10^9 years. They also argue that the existence of such an intercommunicating community is quite likely because, out of the billions of attempts to detect other life that have been made, some will have been successful. A single success, perhaps between two fortuitously close stars, would surely spur both races into further effort at making other contacts, until a vast network was established.

Thus the estimates of L range from 10^2 to 10^9 years. The uncertainty here is greater than in all the astronomical and biological factors. Where does this leave us? Because space has three dimensions, the uncertainty in the range over which we must make contact is much less than the uncertainty in L.

So long as the mean separation between communicative races is less than the thickness of the Galactic disk, this separation varies only as the cube root of $1/L$; for larger mean separation, the variation goes as the square root of $1/L$. In Fig. 1, we have plotted the mean separation S as a function L for three assumptions:

$$\left.\begin{aligned} N &= L \\ N &= \frac{L}{10} \\ N &= \frac{L}{100} \end{aligned}\right\} \text{ corresponding to } \left\{\begin{aligned} F &= 1 \\ F &= 10^{-1} \\ F &= 10^{-2} \end{aligned}\right.$$

The slanting 45° lines give the possible number, n, of two-way exchanges between the races in contact; that is, the lines are drawn for

$$S = \frac{L}{2n}. \tag{6}$$

By a two-way exchange, we mean receiving the response to a transmission at the same end. The round-trip light time does not limit the information rate, only the interaction rate.

We see that if $L < 2000$ years, two-way exchanges are impossible, but that if $L = 10^9$ years, millions of two-way exchanges can occur. Thus, there emerges a rather natural limit to the search range set not only by the thickness of the

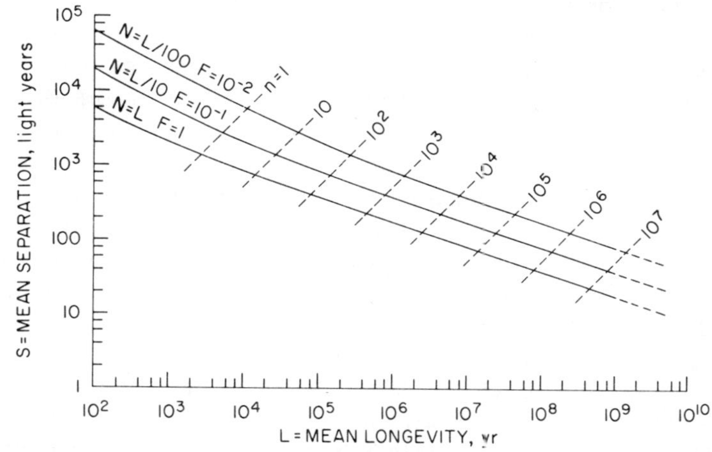

Fig. 1. Relation between N, L and S and the number of possible two-way conversations.

galactic disk in the solar neighborhood, but more forcefully by the relations between longevity and the number of possible exchanges. This limit is on the order of 10^3 light years. Although the maximum range varies only slowly with F and L, the number of stars to be searched varies inversely with both these quantities.

Recently, an analysis was made of the genesis time of intelligence, of the longevity of civilizations, and of the number of coexisting civilizations in the Galaxy (Kreifeldt, 1971). More recently, one of us (Oliver, 1975) examined the probability that coexisting civilizations will arise within a few tens of light years of each other. There are two major conditions under which such circumstances could occur. First, civilizations could coexist on the planetary systems of two separate stars that are fortuitously close to each other. For this case,

$$\nu \sim 10^{-4} L R^3 \tag{7}$$

where ν = total number of civilizations that have been in proximity to others since intelligent life began; L = average longevity of technologically advanced civilizations, years; R = distance between civilizations, light years.

The second case is that of twin civilizations coexisting on the planets of two stars of a binary system. Here

$$\nu_t \sim 10^{-2} L . \tag{8}$$

Adding (7) and (8), we derive our best present estimate of the total number of civilizations that have coexisted within R light years of another:

$$\nu(R) \sim (10^{-2} + 10^{-4} R^3) L . \tag{9}$$

Equations (8) and (9) begin to fail below about 0.1 light year as very widely spaced binaries and multiple systems begin to be excluded.

We are again confronted with the great uncertainty in L. If we solve our problems of planetary management, we could flourish for 10^9 years; if we do not, we may well retrogress or vanish in 10^2 years, as some pessimists fear. It has often been suggested (Shklovskii and Sagan, 1966) that the distribution of longevities may be bimodal; a fraction of the civilizations, say 1%, succeed in solving their ecological and societal problems and live for 10^9 years while the remaining 99% do not and vanish in, say, 10^3 years, on the average. If this is true, then $L \sim 10^7$ years and the Galaxy is teeming with life. Something on the order of 10^5 civilizations will have coexisted within several tens of astronomical units of each other and 10^6 within 10 light years. Interstellar communications will have been established independently countless times.

But let us take the more pessimistic value of $L \sim 10^3$ years, which assumes virtually no long-lived cultures. There will then have been only about 10 twinned civilizations within a few tens of astronomical units of each other. At this close range, the planets of the other star could be resolved and leakage radiation could easily be detected.

For these early radio searches, it is probably more significant to interpret L as the average length of time civilizations radiate, for their own purposes, signals that are detectable at the range R. For a cost less than that for space programs we have already undertaken, we could construct a receiving system that would detect the presence of signals like our own UHF-TV at a 50-light-year range (Oliver and Billingham, 1972). We have been radiating such signals for 15 years and will probably do so for at least another 35 years. Taking $R = 50$ light years and $L = 50$ years, we see that such a strategy could have succeeded about 500 times.

Had we been one of those early "lucky" civilizations, our success would almost certainly have encouraged us to search deeper into space and to radiate a beacon to attract others. Given a high-powered beacon, the situation changes drastically. The same system that could detect leakage radiation at 50 light years could detect a 250-MW beacon at 500 light years and thus reach 1000 times as many stars.

It was also shown (Oliver, 1975) that there are likely to have been about 40 million cases where a space voyage by an advanced culture around one component of a multiple star system to the planets of another component would have discovered either early life evolving, early life that had become extinct, or the remnants of an advanced culture which had not survived. Space voyages over only a few tens of astronomical units are within our near-future capabilities. It is possible also that such discoveries might be made on another planet within the same solar system, further increasing the probability of the discovery of life by a civilization.

The probability that one or more of these discoveries has taken place is high. Such a discovery would provide a strong stimulus to the civilization in question to attempt to contact more distant civilizations, with the evolution of a network of communicating cultures across the Galaxy. It is possible that the rapid increases in knowledge facilitated by such communication would accelerate the evolution of the participants to the point where they are able to achieve stability and longevity. In contrast, those civilizations which do not attempt communication may retrogress. The achievement of interstellar communication, with subsequent membership in the galactic community, could represent a decisive step in the continuation of biological evolution.

Attempts have been made with existing radiotelescopes, over the last 15 years, to detect coherent signals characteristic of intelligent life. Project Ozma, in 1960, was the first. None has been successful to date, which is perhaps not surprising in view of the low sensitivity of the receiving systems and the limited frequency range employed. Most of the activity has been in the microwave region of the spectrum. Some of the searches which are known to have been conducted (Sagan and Drake, 1975) are given in Table 1.

The first attempt to study in detail the design of a system for contacting extraterrestrial intelligent life was made in the summer of 1971 by a team of scientists at the NASA Ames Research Center, under the direction of two of us (Oliver and Billingham, 1973). The concept, known as Project Cyclops, is a large array of phased radiotelescopes on Earth, listening in the low-frequency end of the microwave region of the spectrum for coherent signals from star systems within 1000 light years. This portion of the spectrum was chosen because

Table 1. Attempts to detect signals beamed toward earth by other civilizations

Investigator	Observatory**	Date	Frequency or Wavelength	Targets
Drake	NRAO	1960	1420 MHz	Epsilon Eridani Tau Ceti
Troitsky	Gorky	1968	21 & 30 cm	12 nearby Sun-like stars
Verschuur	NRAO	1972	1420 MHz	10 nearby stars
Troitsky	Eurasian network, Gorky	1972 to present	16, 30, & 50 cm	Pulsed signals from entire sky
Zuckerman Palmer	NRAO	1972 to present	1420 MHz	~600 nearby Sun-like stars
Kardashev	Eurasian network, ICR	1972 to present	Several	Pulsed signals from entire sky
Bridle Feldman	ARO	1974 to present	22.2 GHz	Several nearby stars
Drake Sagan	Arecibo	1975 (in progress)	1420, 1653, & 2380 MHz	Several nearby galaxies

*These attempts have so far been unsuccessful, but the number of stars that have been examined is less than 0.1% of the number that would have to be investigated if there were to be a reasonable statistical chance of discovering one extraterrestrial civilization.

**NRAO is the National Radio Astronomy Observatory in Green Bank, West Virginia; Gorky is the 45-ft antenna at Gorky University in the USSR; Eurasian Network is a network of omnidirectional antennas in the USSR that is being operated jointly by V. S. Troitsky of Gorky University and N. S. Kardashev of the Institute for Cosmic Research (ICR) of the Academy of Sciences of the USSR; ARO is the Algonquin Radio Observatory at Algonquin Park, Canada; Arecibo is the 1000-ft radio-radar antenna at Arecibo Observatory in Puerto Rico.

galactic and quantum noise are at a minimum, atmospheric transmission is good, the deleterious effect of frequency drift rates is minimized, and because the spectral lines of hydrogen and hydroxyl (1400 and 1727 MHz) fall in the center of the region. The data-processing system examines simultaneously all 0.1-Hz channels within this band and can detect signals that are 90 db below the integrated noise level in the band. The phased array would be pointed at each likely star for a period of 20 minutes, and then be switched to the next star. A few million candidate stars are available within 1000 light years, and all could be examined within a few decades by the Cyclops system. The Cyclops system was described in more detail in a paper presented by one of us (Oliver, 1973) at the first CETI Review of the International Astronautical Federation.

In March 1974, the Board of the Scientific Council on the Radio Astronomy Problem Area (1975), of the Soviet Academy of Sciences, considered and approved a research program on the problem of communication with extraterrestrial civilization. The program described is fairly extensive.

During the past decade, ideas concerning the nature of extraterrestrial civilizations have received extensive discussion in the Soviet Union, and searches for pulsed signals from such civilizations have been undertaken at the Gorky Radiophysics Institute, the Institute for Space Research, and the Shternberg Astronomical Institute. The program proposes research into matters relating directly to communication with extraterrestrial intelligent life, including modes of information exchanges between civilizations, deciphering of signals, aspects of the theory of "cosmic civilization," and consequences of contact with other intelligent life. Beyond this, a two-phase observational search program is contemplated. In the first phase, to last until 1985, existing radio-telescopes would observe discrete targets at specific frequencies, while two systems of nondirectional antennas, one ground-based and the other satellite-borne, would continuously monitor the entire sky at all wavelengths between 1 and 100 GHz. In addition, a system of low-directivity antennas of 1° to 3° beamwidth would be used for a continuous survey of nearby galaxies. In the second phase, observations would be conducted by a satellite system continuously monitoring the entire sky and equipped with antennas of large effective area, as well as by a system of two widely spaced ground stations having large (effective area, about 1 km^2) semirotatable antennas for (1) synchronized reception, (2) searches for signals from specific objects, and (3) analysis of selected sources. Finally, a special mountain observatory is contemplated for infrared searches for excessive radiation in that wavelength region from stellar systems which have been altered by astro-engineering projects.

The discussion above refers primarily to questions of searching for signals from extraterrestrial intelligent life. In three different ways, information has been transmitted into space, revealing the presence of our civilization. The first category is the general transmission from terrestrial radio sources over the past 40 years. More recently, engraved plaques were placed on board the Pioneer 10 and 11 spacecraft which conducted the NASA Jupiter flyby missions (Sagan and Drake, 1975). Both spacecraft will leave the Solar System and travel slowly across interstellar distances. The plaques give fundamental information about the planet Earth and the appearance of human beings, in case the spacecraft should ever encounter intelligent extraterrestrials. Finally, in 1974, a message in binary code was transmitted by the radiotelescope at Arecibo toward the Great Cluster in Hercules (Sagan and Drake, 1975). The message can be decoded into 73 groups of 23 characters and read as a picture displaying elementary organic chemistry, genetic code structures, fundamental aspects of human beings and the human population, and details of the Earth and the transmitting telescope. At the wavelength used, 12.6 cm, the Earth will appear nine orders of magnitude brighter than our Sun. This demonstration clearly shows the extraordinary capability now possessed by our civilization for communication over interstellar and inter-galactic distances.

The discovery of extraterrestrial intelligent life could be of profound importance to mankind. Further studies should clearly be carried out. The technology for detection and communication is at hand, and the costs may well be reasonable in comparison with other major endeavors in science, technology, and exploration.

References

Cameron A. G. W. (Editor) (1963) *Interstellar Communication*, Benjamin Press, New York.

Cocconi G. and Morrison P. (1959) Searching for interstellar communications. *Nature* **184**, 844.

Corliss W. R. (1974) *The Viking Missions to Mars*. NASA SP-334.

Haldane J. B. S. (1954) The origins of life. In *New Biology*. No. 16, Penguin, London.

Kaplan S. D. (Editor) (1971) *Extraterrestrial Civilizations—Problems of Interstellar Communication*. NASA TT F-631.

Kreifeldt J. G. (1971) A Formulation for the Number of Communicative Civilizations in the Galaxy. *Icarus* **14**, 419–430.

Lederberg J. (1961) Exobiology: Approaches to life beyond the earth. *Science* **132**, 393–400.

Mallove E. F. and Forward R. L. (1974, 1975) Bibliography of interstellar travel and communication. (I) *J. Brit. Interplanetary Soc.* **27**, 921–943 (1974); (II) *J. Brit. Interplanetary Soc.* **28**, 191–219 (1975); (III) *J. Brit. Interplanetary Soc.* **28**, 405–434.

Miller S. L. (1955) Production of some organic compounds under possible primitive earth conditions. *J. Am. Chem. Soc.* **77**, 2351–2361.

Miller S. L. and Orgel L. E. (1974) *The Origins of Life on Earth*. Prentice-Hall, New York.

Oliver B. M. (1973) State of the art in the detection of extraterrestrial intelligent signals. *Acta Astronautica* **18**, 431–439.

Oliver B. M. (1975) Proximity of galactic civilizations. *Icarus* **25**, 360–367.

Oliver B. M. and Billingham J. (1973) *Project Cyclops: A Design Study of a System for Detecting Extraterrestrial Intelligent Life*. NASA CR-114445 (Revised Edition).

Oparin A. I. (1961) *Life; Its Nature, Origin and Development*, Oliver and Boyd, London.

Ponnamperuma C. and Cameron A. G. W. (Editors) (1974) *Interstellar Communication: Scientific Perspectives*, Houghton-Mifflin, New York.

Sagan C. (Editor) (1973) *Communication with Extraterrestrial Intelligence*. M.I.T. Press, Cambridge, Mass.

Sagan C. and Drake F. (1975) The search for extraterrestrial intelligence. *Scientific American* **232**, 80–89.

Scientific Council on the Radio Astronomy Problem Area, Academy of Sciences of the U.S.S.R. (1975) The CETI program, *Soviet Astronomy* **18**, 669–675.

Shklovskii I. S. and Sagan C. (1966) *Intelligent Life in the Universe*. Holden-Day, New York.

Acta Astronautica. Vol. 6, pp. 59–65. Pergamon Press 1979. Printed in Great Britain

A review of recent concepts of the problem of the origin of life

L. M. MUKHIN
Intercosmos Council, USSR Academy of Sciences, Moscow, U.S.S.R.

Abstract—Conditions suitable for the formation of complicated organic molecules existed in the primitive Earth. Synthesis of such combinations in local regions of the primitive Earth in the conditions far from equilibrium is the most probable one. The main difficulty in understanding the life origin is the formation of genetic code. The hypothesis of the directed panspermia is possible but does not solve the problem of the life origin.

OVER 50 years ago, Oparin (1924) and Haldane (1929) hypothesized a radically new approach to understanding how life originated on Earth. This hypothesis assumes evolution of simple organic compounds formed by the action of various sources of energy on inorganic precursors. Simple organic molecules thus created were available compounds that provided a basis for the formation of more complex substances of biological significance which finally led to the origin of life. The possibility of checking the consequences of this theory experimentally shows the remarkable advantage of the above-mentioned hypothesis.

Starting from the well-known experiments of Miller (1953) to the present, the Oparin–Haldane hypothesis can be validated more and more by experimental investigations. In addition, astronomical data reveal the presence of interstellar molecules that are precursors of complex organic molecules (Winnewisser, 1975).

An essential constraint in solving the problem of the origin of life is that the models of abiogenesis must not contradict chemistry, biology, and planetary evolution theory. The obstacle has already become apparent in attempts to reconstruct the conditions that existed on primitive Earth.

Presently, it is usually assumed that, in the monomer synthesis, the major energy sources were ultraviolet solar radiation, electric discharges, and shock waves. Estimates of the concentrations of organic compounds in the primitive ocean which have been formed by ultraviolet solar radiation cover a wide range. Sagan (1966) assumes that a 1% solution of amino acids and oxi-acids could form in 10^9 years. Abelson (1966) drew attention to the HCN hydrolysis processes that result in the formation of NH_4OH and NCOOH. Assuming that the primitive ocean volume was 10^{22} ml, the author obtained a highly diluted solution of HCN (10^{-4} M); concentration of amino acids would be of the order of 10^{-6} M.

On the other hand, Hull (1960) showed that, considering photochemical destruction processes, stationary concentrations of organic compounds in the primitive ocean would be much lower (10^{-12} M).

The problem of photodestruction of organic molecules in the atmosphere of primitive Earth could be solved by considering vertical transfer in the atmosphere (Mukhin, 1976). If it is assumed that nitrogen and argon are the main components of the primitive atmosphere, and carbon dioxide and water concentrations do not exceed present-day values, then the ozone concentration is about

1% of the current level. Thus, absorption in the region 1700–3000 Å was insignificant and the ultraviolet radiation flux, with $\lambda \sim 3000$ Å, was 10^{15} $cm^{-2}s^{-1}$ and the flux, with $\lambda \sim 1700$ Å, was 5×10^{11} $cm^{-2}s^{-1}$. The energy range of the ultraviolet radiation with these wavelengths is 7.3 to 4.13 eV. Energies of the most important bonds in organic chemistry have the following values (Sagan, 1970): C≡C, 8.4 eV; C=O, 7.4 eV; O–H, 4.8 eV; C–H, 4.3 eV; N–H, 4.1 eV; C–O, 3.7 eV; C=C, 6.4 eV, etc.

Consequently, the ultraviolet radiation of the Sun would have destroyed molecules synthesized by shock waves and electric discharges. The lifetime (from generation to photodestruction) of molecules in the atmosphere can be estimated by order of magnitude as follows: $\tau_{des} \sim 1/\theta\phi\sigma$, where σ is the photodissociation section, ϕ is the photon flux, and θ is quantum efficiency. Then, for aldehydes $\tau_{des} \sim 10^4$ s. The value of the coefficient of eddy diffusion for the lower layers of the atmosphere is $D \sim 4 \times 10^4$ cm^2s^{-1}, with the altitude scale $Z \sim 10$–15 km (Davidson *et al.*, 1966).

In this case, the transfer characteristic time $\tau_{dif} \sim Z^2/D \approx 15 \times 10^7$ s. We see that $\tau_{dif} \gg \tau_{des}$ and the molecule is destroyed before it reaches the ocean's surface. At heights from 15 to 20 km, the coefficient of eddy diffusion is $D \sim 10^3$ cm^2 s^{-1} and, in this layer, the molecule photodissociation conditions are even more significant.

Nevertheless, note that there is some thin layer of the atmosphere where molecules can reach the ocean's surface. The layer thickness can be easily estimated from $Z^2 \leq D\tau_{des}$ and, as a result, $Z \lesssim 10^4$ cm. Bar-Nun and Shaviv (1975) calculated the concentration of amino acids formed by the Strecker mechanism from HCN and HCHO. Hydrocyanic acid and formaldehyde form in the atmosphere (layer thickness, ~5 km) because of shock-wave energy.

Considering the value just calculated for the thickness of the atmospheric layer where the photodestruction processes are insignificant (~100 m), the amino-acid solution equilibrium concentration should be decreased by two orders of magnitude (to 6×10^{-5} M), a concentration that is not sufficient for further molecular evolutionary processes. The above-mentioned values were obtained under the assumption that the system was in equilibrium, which is highly unlikely. Concurrently, one can affirm that thermodynamic equilibrium was never reached on primitive Earth. Synthesis processes occurred to the best advantage in local regions of primitive Earth under conditions far from equilibrium.

What considerations can be advanced on the character of these regions? The analysis of possible conditions for synthesis of organic compounds under the primitive Earth conditions inevitably leads to consideration of volcanic processes. Attention should be paid to some basic aspects:

(1) The lifetimes of active volcanoes and hydrothermal systems are sufficiently long (10^3 to 10^5 years) (Luchitski, 1971).

(2) Volcanic processes constitute a geological factor that has always been present in the history of Earth. There is every reason to believe that, in the early stages of Earth's evolution, the number of active volcanoes was considerably larger than now.

(3) A volcano is a source of energy as well as a source of initial components for the synthesis of organic compounds (water, carbon monoxide, carbon dioxide, methane, ammonia, etc.). If the thermal energy of a volcano required to synthesize monomers is the only value estimated, it is easy to obtain a value of 0.1 cal sec^{-1} cm^{-2} that exceeds, by many orders of magnitude, those for other energy sources with statistical distribution over the Earth's surface.

(4) In an area where there is an underwater volcano or a hydrothermal system, conditions are favorable for the survival of the compounds formed (because of the great pressure and temperature gradients).

(5) The presence of a wide range of natural catalysts in volcanic regions is responsible for further evolution of synthesized compounds.

(6) Photochemical destruction that accompanies underwater ejections and hydrothermal synthesis is weakened substantially.

Significant inequality of the conditions and processes in the regions of volcanic activity should be noted. This circumstance is brought about by the considerable gradients of temperature, pressure, chemical potential, and the processes of organic molecule sorption on mineral matrices.

A recent discovery of HCN (which is a key intermediate in the synthesis of biomolecules), in the hot lava of Alaid volcano, as well as other organic compounds near active volcanoes, supplied strong evidence for this hypothesis (Mukhin, 1974, 1976).

Thus, it is possible that the necessary quantity of the initial organic molecules may have accumulated on primitive Earth. However, a more serious problem concerning the origin of life is the rise and evolution of dynamic living matter.

The Oparin–Haldane view assuming that life arose as a result of a process of producing increasingly more complex compounds, invokes the so-called principle of continuity whereby each evolutionary step can be regarded as a direct consequence of the previous one. Using this principle, one can easily show the possibility of successive formation of rather complex polypeptides from more simple precursors. However, significant difficulties arise when one attempts to understand the origin of the genetic code. This complex process of template synthesis is the result of the long period of evolution, and the stages of its development appear to be lost.

The universal character of the genetic code is best explained if there is a stereochemical correspondence between amino acids and base triplets. This view was developed by Pelc and Welton (1966) and Welton and Pelc (1966), but Woese (1966), after analyzing their arguments, does not agree with their conclusions. Crick (1968) emphasizes that, even if such correspondence is possible, great difficulties would be encountered in explaining the mechanism of ribosomal translation.

To understand the evolution of the genetic code, interaction between protein and nucleic acid precursors, as well as between irregular polypeptides and polynucleotides, is required and the shift from the abiogenically synthesized molecules to a dynamic organization with initial elements of transcription and translation would be of primary interest.

Of special importance is the observation that, in all experiments on abiogenic synthesis, the yield of nitrogen bases is always lower than that of amino acids, and thymine, guanine, and cytosine are produced in extremely small quantities. Moreover, conditions optimal for the synthesis of one base are not optimal for the synthesis of any other. While a family of amino acids appears under specific conditions, bases are formed as "singlets." Note, in addition, that while the amino-acid polycondensation is a rather simple process, similar experiments with nucleotides were less successful. This is still a major problem in our understanding of the origin of life (Orgel, 1968). Assume that there exists some pool of amino acids, polypeptides, bases and nucleotides, and consider the possible evolutionary pathways.

It is usually believed that most problems connected with the origin of life can be solved if one assumes that primary polypeptides had a weak catalytic function and primary polynucleotides had a weak replicative function. We wish to emphasize that the proposal of the catalytic activity of polypeptides seems more convincing. In this case, a primary polypeptide could perform functions similar to those of DNA polymerase. Since proteins are probably "older" than nucleic acids, this hypothesis could be tested experimentally. Further, the formation of polynucleotides would have initiated the beginning of template synthesis of proteins, suggesting that an autocatalytic process may have started.

It is very tempting to think that the polypeptides first appeared as templates for the nucleotide condensation and only at a later date did the replicative function arise, possibly as late as the formation stage of protocells. One may suggest the following argument for this. The contemporary processes of genetic replication and protein template synthesis occur in solutions with a very high concentration of organic substances. It is unlikely that there were regions on primitive Earth with such high a concentration. Most probably, the system of coupled reactions and the mechanism of replication and synthesis both occurred in concentrated solutions only. Not until some critical concentration (like a critical mass in nuclear physics) was reached did the replicative processes begin. It is possible that the critical concentration may have been attained with protocells (coacervates or microspheres), which had the ability to concentrate material.

Thus, we suggest that the replicative function arose later than the protocell, and it was preceded by a prolonged evolution of polymers and a number of morphological entities (protocells). Of course, this simplified scheme does not explain how the control of DNA replication in a cell is achieved, how this control mechanism appeared, or the reasons for the surprising reliability of the copying mechanism.

To show that this is not too speculative, we shall discuss various experimental data (Fox and Dose, 1972). Most attempts to reveal some interaction between amino acids and polynucleotides have been unsuccessful. Much more successful are the experiments on the interaction between polyaminoacids and mononucleotides (Woese, 1968). The strength of the interaction of polyarginine with bases was shown to make a series: $G < A < C < U$. The same order was seen for interactions with polylysine. Thus, a possibility exists that polypeptides can function as a template at the initial stages of evolution (the inverse of translation).

An interesting interaction of thermally synthesized proteinoids of a morphologically distinct structure with polypurines and polypryimadines has been studied in Fox's laboratory. It was established that one of the proteinoids studied had different chemical affinities for different polynucleotides (Fox *et al.*, 1971).

Undoubtedly, experiments of the greatest interest are those that model the evolution of protein biosynthesis (and thereby the evolution of genetic apparatus). Recently, Nakashima and Fox (1972) investigated the incorporation of amino acids (as aminoacyladenylates) into microparticles composed of a lysine-rich proteinoid and polyribonucleotides. They demonstrated a selectivity on the part of polyribonucleotides toward incorporation of an amino acid. They proposed a mechanism of polypeptide synthesis which preceded contemporary translation. Their proposal is based, in particular, on experiments which showed that the 2′–OH group poly-A and poly-U could be acetilated (70–80%) with aceticanhydride. In other experiments in which acetylaminoacyl imidazols were the activated form of amino acids, it was demonstrated that an ester bond with 2′–OH groups could form.

These data suggest a preribosomal mechanism of translation which uses the 2′–OH groups of the sugar-phosphate backbone of RNA.

Certainly, these experiments should be regarded as the first steps in the new field of prebiological synthesis, the modeling of dynamic processes. A major difficulty is that the conditions of the "test-tube" experiments (concentrations of reagents, chemical reactions used, etc.) are inconsistent with geological and geochemical conditions that existed on primitive Earth about 4×10^9 years ago (although local environments can be invoked). Nonetheless, we think that experimental and theoretical work in this field is of principal importance; it may well be the field in which significant achievements in our understanding of the origin of life can be expected.

Lippman's experiments on gramicidine biosynthesis under direction of enzymes in a system—amino acids—intermediate peptides—also provide evidence for a primary role of polypeptides instead of nucleic acids. However, experimental evidence of a protein-directed copying system is absent. True, some data on specific interactions of mononucleotides with polyaminoacids have been obtained, but it should be remembered that synthesis of polynucleotides on polypeptide templates has not yet been demonstrated.

A major problem arose in our understanding of the origin of life with the creation of Eigen's hypercycle (Eigen, 1971), i.e. of a closed system with all the control and executive mechanisms necessary for replication. The difficulties were so tremendous that ideas of "extraterrestrial" origin of the "terrestrial" life were suggested.

Recently, Crick and Orgel (1973), taking into account cosmological data, proposed a revised model of panspermia—the so-called "directed" panspermia—which explains the origin of life on Earth by the purposeful activity of civilizations older than our Sun. Until we understand the genetic apparatus we cannot completely reject the idea of panspermia, regardless of the suggested form. Yet one must realize that this idea does not solve the problem of the origin of life, but rather transfers it to another place and another time. The Crick–Orgel hypothesis is that there could be other planets on which the conception of life might be more probable. On such planets, minerals present only in low concentrations on Earth and whose catalytic activities are especially important for the origin of living systems would be more abundant. Furthermore, it is important that such planets predate the formation of the Earth.

The age of the Galaxy is believed to be 13×10^9 years, and during the 2×10^9 years after the Galaxy formed, a star containing both light and heavy elements might have formed. Thus, 6.5×10^9 years before the formation of our Solar System, a planetary system may have appeared with a planet of the Earth-type. Since the time of origination of technological civilization on Earth is less than 4×10^9 years, technological civilizations might have existed in the Galaxy before Earth's formation.

Proceeding from these astrophysical data, Crick and Orgel (1973) suggest that life on Earth might have originated as a result of infection of our planet with some microorganisms specially sent by a highly developed civilization to "spermatize" planets potentially suitable for life. Considering the immediate possibilities of our own civilization, one can well see that the construction of a such a spaceship presents a real task. In defense of their hypothesis, Crick and Orgel forwarded the following arguments:

The chemical content of the living systems must reflect (though not absolutely) the conditions which existed at the moment of their origin. The presence in terrestrial life of rare elements may therefore testify to the extraterrestrial origin of our life. The content of chromium, nickel, and molybdenum on Earth is 0.2, 3.16 and 0.02%, respectively. And while the first two elements play practically no role in biochemistry, molybdenum is required for many enzymic reactions. If it were possible to prove that elements present in the Earth's organisms correlate with the abundances of the same elements in the so-called "molybdenum stars", a good argument for the above hypothesis could be made.

A much more serious argument in favor of the directed panspermia hypothesis comes from the universal character of the genetic code. Since, at present, there is no satisfactory theory on the origin and universal character of the code, the hypothesis under consideration is the only one that easily solves this difficulty.

Note that, while recent publications (Hoffmann, 1975; Wong, 1976) suggest new approaches to our understanding of the evolution of the genetic apparatus, the question of code correspondence remains unanswered. This brief analysis shows that the problem of the origin and evolution of the genetic code is full of unknown facts and far from a final solution.

Acknowledgement—I thank Dr. L. Minchenkova, Dr. V. I. Ivanov and Dr. Frank-Kamenetsky for useful discussions.

References

Abelson P. H. (1966) *Proc. Nat. Acad. Sci.* **55**, 1365.
Bar-Nun A. and Shaviv A. (1975) *Icarus* **24**, 197–210.
Crick F. H. C. (1968) *Mol. Biol.* **38**, 367.

Crick F. H. C. and Orgel L. E. (1973) *Icarus* **19**, 341.
Davidson *et al.* (1966) *Tellus* **18**, 3a.
Eigen M. (1971) *Naturwis.* **58**, 465.
Fox S. W. and Dose K. (1972) *Molecular Evolution and the Origin of Life.* W. H. Freeman, San Francisco.
Fox S. W. *et al.* (1971) In *Nucleic Acid-Protein Interaction* (Edited by D. Ribbons and F. Woessner), p. 113. North-Holland, Amsterdam.
Haldane J. B. S. (1929) *The Origin of Life.* Rationalist Annual.
Hoffmann G. W. (1975) *Ann. Rev. Phys. Chem.* **26**, 123–144.
Hull D. E. (1960) *Nature* **186**, 693.
Luchitski I. V. (1971) *Advances in Paleovolcanology.* Nauka. Moscow.
Miller S. L. (1953) *Science* **117**, 528–529.
Mukhin L. M. (1974) *Nature* **251**, 50.
Mukhin L. M. (1976) *Origins of Life* **7**, 335–368.
Nakashima T. and Fox S. W. (1972) *Proc. Nat. Acad. Sci. U.S.A.* **69**, 106.
Oparin A. J. (1924) *The Origin of Life* (Translated from the Russian and reprinted in Bernal J. D. (1967) *The Origin of Life*, pp. 199–241, Weidenfeld and Nicholson, London.)
Orgel L. E. (1968) *J. Mol. Biol.* **38**, 381.
Pelc S. R. and Welton M. G. E. (1966) *Nature* **209**, 868.
Sagan C. (1966) In *The Origin of Prebiological Systems.* Mir, Moscow.
Sagan C. (1970) *Encyclopedia Britannica*, pp. 1083–1087.
Welton M. G. E. and Pelc S. R. (1966) *Nature* **209**, 870.
Winnewisser G. (1975) *Naturwis.* **62**, 200–210.
Woese C. R. (1966) *Cold Spring Harbor Symp. Quant. Biol.* **31**, 723.
Woese C. R. (1968) *Proc. Nat. Acad. Sci. U.S.A.* **59**, 110.
Wong J. T. (1976) *Proc. Nat. Acad. Sci. U.S.A.* **73**(7), 2336.

Acta Astronautica. Vol. 6, pp. 67–69. Pergamon Press 1979. Printed in Great Britain

An extended Drake's equation, the longevity–separation relation, equilibrium, inhomogeneities and chain formation

R. N. BRACEWELL
Stanford University, Stanford, CA 94305, U.S.A.

Abstract—Drake's equation implies that the number of galactic communities technologically more advanced than our own is proportional to the average longevity of a community. But if there are significantly different kinds of community such a simple equation may not be adequate for estimating numerical values. Extra terms are needed for each different kind. Spatial inhomogeneity of coexisting communities is also critical and is not reflected by a single universal average. Information flow resulting from contacts between accidentally close-spaced communities could be important in extending life to those communities. Spatial inhomogenity or other factors leading to temporal nonequilibrium could be dominant in determining the number of superior galactic communities exant today. The plus-or-minus on estimates of population by Drake's equation spans a number of cases where different courses of action on our part are implied. It is therefore unavoidable in practice to identify the various regimes case by case and to consider corresponding plans for action.

WITH A VIEW to estimating the number of intelligent, or rather *communicative*, communities in the Galaxy, Frank D. Drake introduced an equation which, in abbreviated form, may be written

$$N = RL \tag{1}$$

where N is the number of galactic communities "at or beyond our level of technological advance" (Sagan, 1973b) and thus able to communicate, R is their rate of appearance, and L is their average longevity to the time when they fall below our current level. By separately estimating R and L, one hopes to arrive at an estimate of N.

The quantity R is an abbreviation for numerous factors such as the rate of star formation, the number of habitable planets per star, and the fraction of habitable planets where communicative life has in fact developed. Various authors have chosen the factors in various ways.

For a preliminary discussion, eqn (1) offers a reasonable point of departure, but it limits attention to just one way in which communicative life might arise, namely, the way in which we think man's Solar System produced man. Drake's equation is an oversimplification.

One would not attempt to estimate N_f, the number of fish in the sea, by multiplying an estimate of R_f, the rate of formation of fish, by L_f, the average longevity of a fish. There are too many ***kinds*** of fish to make this simplification ***useful***. At the least, additional terms allowing for different paths to communicative life are necessary.

I suggest, as a minimum refinement, a modified equation:

$$N = \sum_{j=1}^{J} R_j L_j \tag{2}$$

a sum of terms which admits that a jth kind of communicative society may be arising at a rate R_j under conditions where the associated longevity L_j leads to a current number of members $R_j L_j$ that may be a significant fraction of the total number N even though $R_j \ll R$ or $L_j \ll L$.

Serious proposals contemplating two species of life ($J = 2$) have been put forward by Sagan (1973a,b) and Billingham *et al.* (1975). Another serious proposal involving three levels of existence was that "in a galaxy supporting only 10^3 communities with brief expectation of life, there may be some communities that have achieved durability, even quasipermanence, perhaps by gaining control of the circumstances that lead to short average lifetimes. Aided by accidental proximity due to random spacing, some of these could be in contact" (Bracewell, 1960).

Calculations based on eqn (1) have led to estimates of N suggesting that it may be hundreds of light years or more to the nearest more advanced community. But if there were two levels of existence, one comprising *newborn* communities subject to whatever average longevity applies to such communities and a second level of *durable* communities, the circumstances could be different. For one thing, the concept of average longevity is defeated, and we see the second level gaining in population in a way depending on the age of the universe. Equilibrium statistics no longer apply.

If now, in addition, there is a third level of *communities in contact*, or a galactic club, even more sophisticated possibilities need to be considered. For example, transfer of information between communities in contact could influence not only their survival (von Hoerner, 1961), but also lead to the spawning of new communities. New communities might occupy habitable planets but, from our present standpoint of a search for extraterrestrial communicators, we would need to entertain the possibilities of space-dwelling colonies and of automatic quasi-intelligent probes (Bracewell, 1976).

Other refinements of Drake's equation are needed to take account of spatial and temporal inhomogeneity in all the various parameters. Even at low population densities, accidents of proximity, as shown by Oliver (1975), can enable contacts to occur that would seem unlikely if only the *mean* spacing of communities were considered.

The net effect of the refinements is to indicate that the totality of intelligent life in our Galaxy is likely to exhibit evolution in time. We should broaden our vision to include new parameters such as the ratio of the number of links in a galactic chain of contacts to the number of communities now able to communicate but facing an uncertain future. As time elapses, this ratio should vary. I do not think this ratio can be estimated numerically, any more than we can estimate N from Drake's equation; I would prefer to divide up the full range of possibilities according to the operational consequences. Possibly the ratio is now much less than unity, but caution is indicated.

It cannot be said for certain that the number of club members is small because, for one thing, the numerous cool M-dwarf stars have not convincingly been ruled out as sites for life, original or colonial. For that reason, a radio search for extraterrestrial intelligence should allocate substantial listening time to directions in space other than those directions where we see Sun-like stars. The possible motivations of club members should be a matter for study because the actions of a club are more likely to be extroverted than are those of a solitary community. For that reason, too, chain formation must be incorporated as an essential feature of the basic equations.

References

Billingham J., Oliver B. M. and Wolfe J. H. (1975) *A review of the theory of interstellar communication.* International Astronautical Federation, Lisbon, IAF-A-75-31.

Bracewell R. N. (1960) Communications from superior galactic communities. *Nature* **186**, 670.

Bracewell R. N. (1976) *The Galactic Club.* W. W. Norton, New York.

Oliver B. M. (1975) Proximity of galactic civilizations. *Icarus* **25**, 360–367.

Sagan C. (Editor) (1973a) *Communication with Extraterrestrial Intelligence* (*CETI*). M.I.T. Press, Cambridge, Mass.

Sagan C. (1973b) On the detectivity of advanced galactic civilizations. *Icarus* **19**, 350–352.

von Hoerner S. (1961) The search for signals from other civilizations. *Science* **134**, 1839.

Acta Astronautica. Vol. 6, pp. 71–79. Pergamon Press 1979. Printed in Great Britain

Rationale for the water hole

B. M. OLIVER†

Hewlett–Packard Company, Palo Alto, CA 94304, U.S.A.

Abstract—A band of frequencies extending from 1400 to 1700 MHz and embracing the hydrogen and hydroxyl spectral lines has been dubbed the "water hole" and is considered to be a prime band for the search for extraterrestrial intelligence (SETI). This paper gives the physical and psychological basis for this belief.

Introduction

TWO DECADES ago, Cocconi and Morrison (1959) suggested that we search at frequencies near the hydrogen line for signals emitted by advanced extraterrestrial civilizations attempting to establish contact with us. At the time, the hydrogen line was believed to be unique, but, since then, dozens of other microwave emission lines from a wide variety of interstellar molecules have been discovered. In 1971, the Cyclops study (Oliver and Billingham, 1973), for reasons that are believed to be rather fundamental, identified the band between 1400 and 1700 MHz bounded at the low end by the hydrogen line (1420 MHz) and at the high end by the hydroxyl lines (1662+ MHz) as the prime region of the spectrum to be searched for interstellar signals. Because of these limiting markers, the Cyclops team dubbed this region the water hole and suggested that different galactic species might meet there just as different terrestrial species have always met at more mundane water holes.

At present there is no serious interference in the water hole, but navigational satellites and other systems are being planned that would fill the band with interfering signals such as continuous pseudo-random wide-band noise. If these systems become operational as allocated, a substantial fraction, if not all, of the water hole will be rendered unusable for the search. The proposed services can be shifted to other frequencies without appreciable loss of effectiveness but, if the rationale for the water hole is correct, the search for intelligent extraterrestrial life cannot. It would be a bitter irony if the desire to know exactly where we are at all times on Earth were to prevent us from ever knowing where we are with respect to other life in the Galaxy. It is therefore timely to reexamine the case for the water hole in order that we do not, out of ignorance or carelessness, forever blind ourselves to the signals from advanced societies.

The basic premise that leads us to the water hole is that any advanced society wishing to establish contact will choose the least expensive means that will nevertheless assure success. As we shall see, one of the dominating factors is the *energy* that must be expended by the society to announce its existence over interstellar distances, not just to us but to all likely planetary systems. It is this consideration that leads to the radiation of electromagnetic waves rather than probes or spaceships and to the spectral region of the water hole.

†Vice President for Research and Development.

Case for electromagnetic waves

In all probability we will have to examine thousands of targets before we succeed in detecting intelligent life. The nearest civilization is probably several tens and maybe hundreds of light years from the Sun. With such a profusion of targets, all at such enormous distances, it appears that search by actual manned space travel or by sending out swarms of probes is out of the question. This approach is far too consumptive of time and energy.

Just to send *one* spaceship to a nearby star and return it in twice the round-trip light time, using not fusion or fission power with low yields but matter-antimatter annihilation, would require $\sim 10^{24}$ J. This is enough energy to supply the present total U.S. electrical power for 50,000 years or to keep a 1000-MW omnidirectional beacon shining for 10 million years. Such a beacon could be received by civilizations around any of the million or more good suns within 1000 light years, and would probably not need to be radiated by the searching society for more than 10^3 years. If so, the beacon is 10^{10} times as effective per joule.

If we seek to reduce the energy requirements of space travel by, say, 10^4, we incur round-trip travel times of 400 years per light year or 1600 years for the nearest star. Clearly, we are not going to find other intelligent life by hurling tons of matter through space but by receiving—and possibly some day sending—some form of radiation.

Regardless of what form of radiation is used, in order to detect that a signal exists and is of artificial origin:

(a) The number of particles received must significantly exceed the natural background count.

(b) The signal must exhibit some property not found in natural radiations.

In addition, the radiation should:

(c) require the least radiated power,
(d) not be absorbed by the interstellar medium or atmospheres,
(e) not be deflected by galactic fields,
(f) be readily collected over a large area,
(g) permit efficient generation and detection,
(h) travel at high speed, and
(i) normally be radiated by technological civilizations.

Requirements (a), (c) and (h) taken together virtually exclude from consideration all particles except those having zero rest mass. The kinetic energy of an electron traveling at half the speed of light is 10^8 times the total energy of a 150-GHz photon. All other factors being equal, the electron communication system would require 100 million times as much power. Baryons are worse. In addition, all charged particles fail requirements (d) and (e). Of the zero rest mass particles, gravitons and neutrinos fail requirements (f), (g) and (i). Of all known particles, only low-energy photons meet all the listed criteria. It is almost certain that interstellar communication, if it exists, is accomplished by electromagnetic waves.

Requirement (b) is satisfied by any spatially and temporally coherent electromagnetic wave. Modulation of such a wave in order to transmit information need not destroy the coherence to such an extent that it would be mistaken for a natural signal. Further, the modulation will very likely contain regularities and repetitions of complex patterns that are inexplicable from natural sources.

Optimum spectral region

Electromagnetic radiation covers a practically unlimited frequency range. Unless we can find some reason to prefer a relatively narrow portion of the spectrum, one of the dimensions of the search space remains unbounded. Here again we can appeal to energy minimization.

Let us examine critically what is implied by requirement (a) above. First let us assume that we are operating in the "quantum" region of the spectrum (i.e. $h\nu/kT > 1$), where direct photon detection is easy, and that there is no background radiation. In principle, we require at least one received photon in the observing time τ to detect a signal at all. This requires that the received power

$$P \geq \frac{h\nu}{\tau} \tag{1}$$

so, for a constant collecting area, the equivalent isotropic radiated power (EIRP) must be proportional to frequency. In practice, the reception of a single photon would hardly convince us that we had detected extraterrestrial intelligence. To determine the coherence and other properties of the signal we would need to receive some number n of photons. But n depends only on the sophistication of our data processing and on our assigned *a priori* improbability and does not depend on operating frequency. Thus the proportionality implied by (1) still holds as long as there is no background.

If the radiated signal is coherent and of constant amplitude, the photon arrivals will be Poisson distributed. If the expected number of arrivals in the observing time τ is $\bar{n}$, the mean square fluctuation will also be $\bar{n}$. The detected signal/noise *power* ratio (DSNR) is the ratio of the square of the mean to the mean square fluctuations, i.e.,

$$\text{DSNR} = \frac{(\bar{n})^2}{\bar{n}} = \bar{n} = \frac{W}{h\nu} = \frac{P\tau}{h\nu}. \tag{2}$$

In communication systems, τ is the Nyquist interval and is the reciprocal of the RF bandwidth B. Then (2) becomes

$$\text{DSNR} = \frac{P}{h\nu B} \tag{3}$$

and the photon shot noise behaves as if it had a spectral power density $h\nu$.

The above considerations suggest that the lower the operating frequency the less will be the required EIRP; this is true until we reach the "thermal" region (i.e. $h\nu/kT < 1$). In this region, direct photon detection becomes difficult but the "radio" techniques of linear amplification, mixing, and spectral analysis are applicable; in fact, these have by now been extended into the optical region. Suppose we Fourier transform a block of the received IF signal of duration τ. In effect, a constant-amplitude monochromatic signal is then detected as a rectangular pulse of duration τ to which noise has been added. The effective selectivity curve of the Fourier transform will be the transform of the time window or

$$F(\omega) = \frac{\sin(\omega - \omega_0)\tau/2}{(\omega - \omega_0)\tau/2} \tag{4}$$

and this is the matched filter through which a CW pulse of frequency ω_0 and duration τ should be passed to give the greatest ratio of peak output signal power to mean-square fluctuation in the peak. For this case or any equivalent optimum detection process, the detected signal/noise ratio is

$$\text{DSNR} = \frac{W}{\Psi} = \frac{P}{\Psi B} \tag{5}$$

where Ψ is the spectral noise power density.

Three contributions to Ψ are unavoidable and they are essentially the same anywhere in the Galaxy. These are shown in Fig. 1 (divided by k to give their equivalent noise temperature). The first is the synchrotron radiation of the Galaxy itself given by

$$\Psi_g \sim kT_0(10^9/\nu)^{2.5} \tag{6}$$

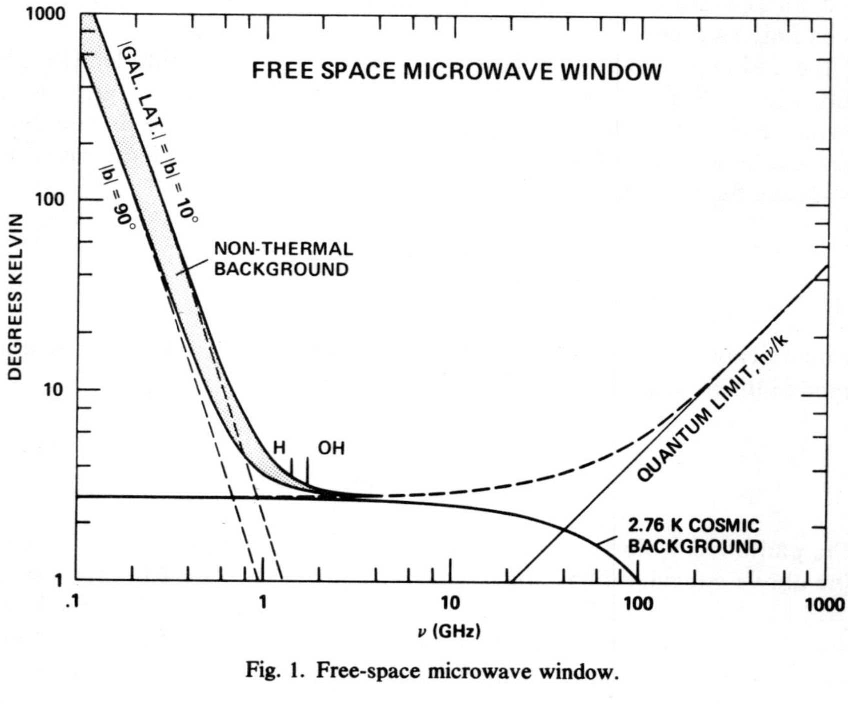

Fig. 1. Free-space microwave window.

where the coefficient T_0 varies from about 1 to 2.5 depending on galactic latitude. This noise dominates below 1 GHz but, above this frequency, rapidly becomes less than the relict cosmic background radiation:

$$\Psi_b = \frac{h\nu}{e^{h\nu/kT} - 1} \tag{7}$$

where $T \sim 2.76°K$. Finally, for $\nu > kT/h \sim 60$ GHz, the spontaneous emission noise

$$\Psi_0 = h\nu \tag{8}$$

becomes dominant.

The sum of these noise contributions defines a broad quiet region extending from 1 to 60 GHz known as the *free-space microwave window*. On Earth (or on any Earth-like planet), the absorption lines of the water vapor and oxygen in the atmosphere re-radiate noise with broad peaks at 22 and 60 GHz which degrade the window above about 10 GHz (as shown in Fig. 2). Thus the *terrestrial microwave window* extends from ~1 to ~10 GHz, and is clearly a preferred region of the spectrum for interstellar search using low noise receivers.

The suggestion is sometimes made that one might avoid ψ_g and ψ_b by going to very high frequencies and then eliminate Ψ_0 by employing direct photon detec-

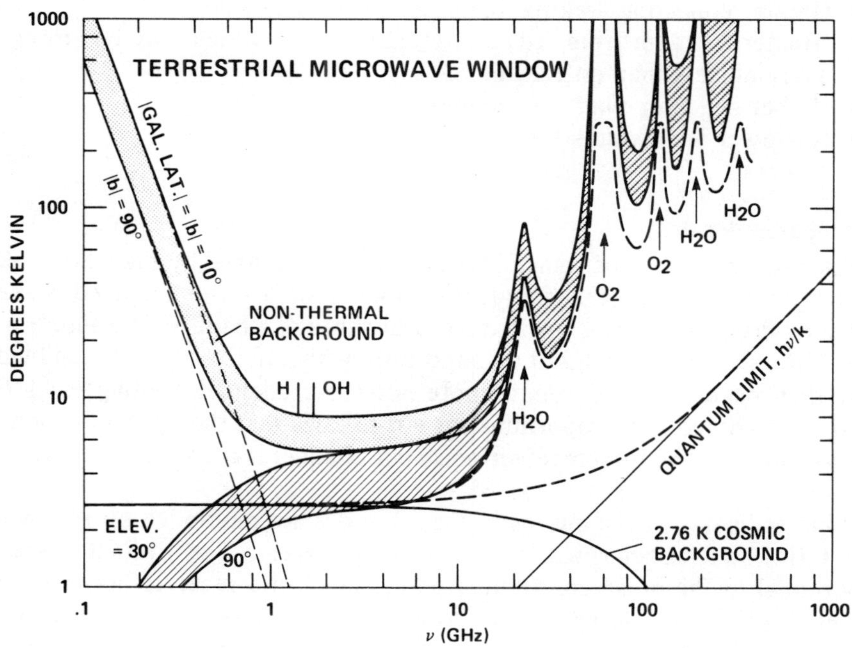

Fig. 2. Terrestrial microwave window.

tion. At frequencies for which Ψ_0 dominates, (5) becomes

$$\mathrm{DSNR} = \frac{P}{h\nu B} \tag{9}$$

and, since B can, in principle, be made as narrow as $1/\tau$, we find

$$\mathrm{DSNR} = \frac{P\tau}{h\nu} \tag{10}$$

exactly as given by (2). Thus ignoring technological limitations, linear amplification is just as good as direct photon detection. Spontaneous emission noise does not make linear amplifiers noisier than photon detectors, it merely prevents them from being quieter, which would violate the principle of complementarity (Heffner, 1962).

It is true that we may be unable, at optical frequencies, to realize bandwidths as narrow as $1/\tau$. The point is that, even if we could, the linear amplifier, which would then be no noisier than the photon detector, would still be far noisier than a linear amplifier operating in the microwave window.

Not only is the minimum detectable received power least in the microwave window, but also the cost per unit of collecting area is less there than for higher frequencies. This latter consideration also favors the low end of the microwave window over the high end. Additional factors favoring the low end are:

- Greater collecting area for the sharpest usable beam.
- Greater freedom from H_2O and O_2 absorption, which may be stronger on many planets than on Earth.
- Higher beacon powers are probably easier to achieve.
- Narrower bandwidths are possible.

Effect of Doppler drift

The relative motion of transmitter and receiver caused by planetary rotation and revolution produces a frequency modulation of the received signal. A frequency drifting at a rate $\dot{\nu}$ will drift clear through a band B in a time $\tau = B/\dot{\nu}$. If $\tau < 1/B$, the receiver will not respond fully while, if $\tau > 1/B$, the bandwidth is larger than necessary and admits more noise per channel. Setting $\tau = 1/B$, we find $B = \dot{\nu}^{1/2}$. Since $\dot{\nu}$ is proportional to ν it follows that the optimum bandwidth increases as $\nu^{1/2}$. The total receiver noise per channel is therefore proportional to $\nu^{1/2}\Psi(\nu)$.

If the ordinates of the curves in Figs. 1 and 2 are multiplied by $\nu^{1/2}$ (where ν is in GHz), we obtain Figs. 3 and 4, respectively. The ordinates are now proportional to the noise contributed per channel in a receiver optimized at all frequencies for the Doppler rate produced by a given radial (line-of-sight) acceleration. We see that the upper end of the free-space window has become relatively noisy and that the H and OH lines are at the quietest part of the spectrum.

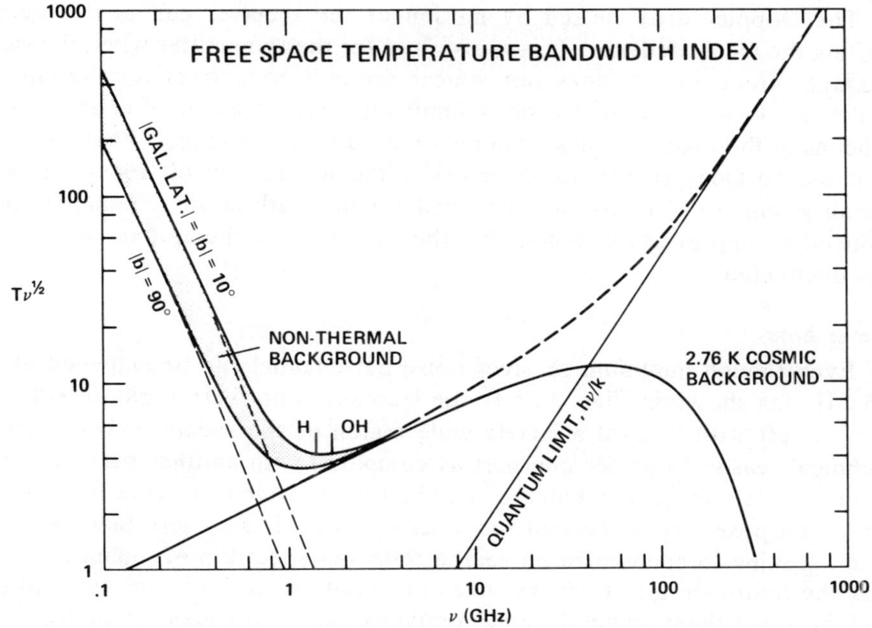

Fig. 3. Free-space temperature bandwidth index.

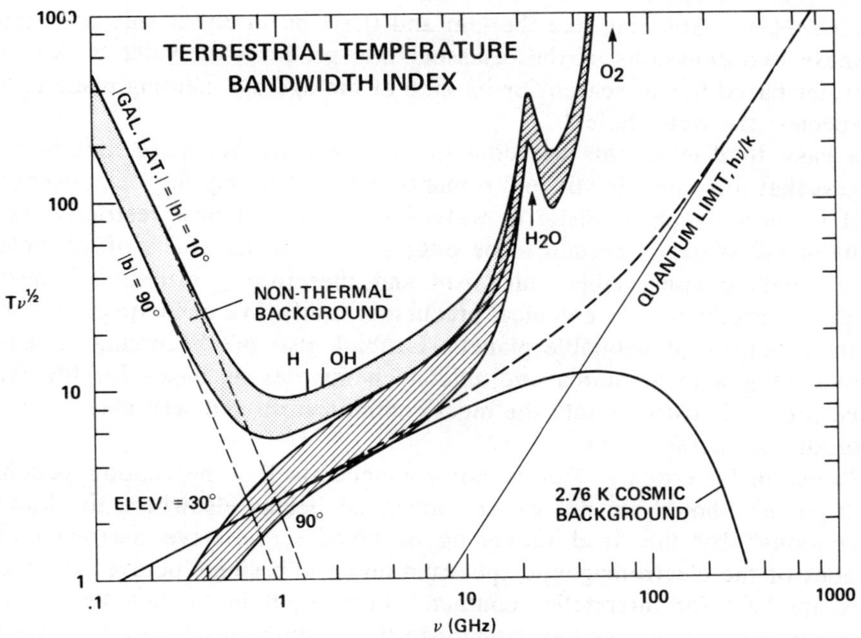

Fig. 4. Terrestrial temperature bandwidth index.

The Doppler drift caused by motion of the receiver can be corrected by drifting the local oscillator. This is also true for the transmitter when the signal is radiated directively. It does not appear possible to correct a coherent omnidirectional beacon for all directions simultaneously. One can, of course, put such beacons at the planet's poles or in long period orbits in space. Whatever is done to minimize Doppler and oscillator drifts, the square root of frequency proportionality remains. The absolute achievable bandwidth at any frequency may be reduced by improved technology but the frequency of the minima in Figs. 3 and 4 is unaffected.

Water hole

Even though minimum receiver noise per channel can be achieved at about 1.5 GHz (as shown in Figs. 3 or 4), the increase is not very rapid on either side. We are left with at least a 2-GHz-wide region of the spectrum with very little technical reason to prefer one part as compared with another part. Not only is this too wide a frequency band to search, it is too wide to reserve for interstellar search purposes, being needed for other services. It was here that the Cyclops team, seeking to economize on search time and spectrum occupancy, observed that the hydrogen and hydroxyl lines are right at the optimum spectral region and, between them, define a rather ominous band. Conclusion 6 of the Cyclops report reads:

> "Nature has provided us with a rather narrow band in this best part of the spectrum that seems especially marked for interstellar contact. It lies between the spectral lines of hydrogen (1420 MHz) and hydroxyl radical (1662 MHz). Standing like the Om and the Um on either side of a gate, these two emissions of the disassociation products of water beckon all water-based life to search for its kind at the age-old meeting place of all species: the water hole."

It is easy to dismiss this as romantic, chauvinistic nonsense, but is it? We suggest that it is chauvinistic and romantic but that it may not be nonsense.

It is certainly chauvinistic to water-based life, but how restrictive is such chauvinism? Water is certain to be outgassed from the crusts of all terrestrial planets having appreciable vulcanism and therefore a primitive atmosphere capable of producing the chemical precursors to life. We can expect seas to be a common feature of habitable planets. Exobiologists are becoming increasingly disenchanted with ammonia and silicon chemistries as bases for life. Water-based life is almost certainly the most common form and well may be the only (naturally occurring) form.

Romantic? Certainly. But is not romance itself a perception peculiar to intelligence? Should we not expect advanced beings elsewhere to show such perceptions? By the dead reckoning of physics, we have narrowed all the decades of the electromagnetic spectrum down to a single octave where conditions are best for interstellar contact. There, right in the middle, stand two signposts that, taken together, symbolize the medium in which all life we know began. Is it sensible not to heed such signposts? To say, in effect, I do not trust your message, it is too good to be true!

In the absence of any more cogent reason to prefer another frequency band, we suggest that the water hole be considered the primary preferred frequency band for interstellar search. This does not mean that other frequencies should be ignored. Harmonics of the hydrogen line deserve some attention. In space, the waterline itself, at 22 GHz, may have merit. It does mean, however, that the water hole deserves the greatest attention for protection against interference.

It is always possible to dismiss the above argument on the grounds that we do not know everything yet and that there may be some more compelling reason to choose another frequency band or even some as-yet-to-us unknown method of communication. These assertions are undeniable and also unacceptable since they logically lead to never doing anything. If we are to make progress we must proceed on the basis of what we know, and not forever wait for something now unknown to be discovered.

References

Cocconi G. and Morrison P. (1959) Searching for interstellar communication. *Nature* **184**, 844–846.

Heffner H. (1962) The fundamental noise limit of linear amplifiers. *Proc. IRE* **50**, 1604–1608.

Oliver B. M. and Billingham J. (1973, Rev. Edition) *Project Cyclops: A Design Study of a System for Detecting Extraterrestrial Intelligent Life*, NASA CR 114445.

Acta Astronautica. Vol. 6, pp. 81–94. Pergamon Press 1979. Printed in Great Britain

Search for radio emissions from extraterrestrial civilizations

V. S. TROITSKII, A. M. STARODUBTSEV AND L. N. BONDAR

Radiophysical Research Institute, Gorki, U.S.S.R.

Abstract—Systematic observations of the whole celestial halfsphere radiation at centimeter and decimeter wavelength ranges have been carried out since 1970. Simultaneous observations at four locations in the Soviet Union are made to filter out cosmic signals from a background of all kinds of local noise. The existence of a global sporadic radiation apparently arising near the Earth was discovered. Conclusions are reached concerning the optimal conditions for scanning the cosmos in order to detect sporadic radiation from technically developed extraterrestrial civilizations.

WE BEGAN our study by searching for sporadic radio emissions from space arising as a result of engineering activity of an extraterrestrial civilization possibly located around one of the stars nearest the Sun.

Intelligent life must exist in the Universe not only near our star, the Sun. A regular evolutionary property of matter leads inevitably to the development of intelligence. The search for intelligence in the Universe, via messages sent long ago by extraterrestrial civilizations, is already in experimental study programs. A search for messages possibly sent to Earth from nearby stars was carried out by F. Drake (U.S.A.) in 1961 and then more extensive investigations involving nearly 10 of the nearest stars were made at NIRFI (Gorki) in 1968. In 1973 Verschuur (U.S.A.) again searched for signals from some of the nearest stars.

An extraterrestrial civilization, however, may not transmit messages directed to other stars such as ours. The most effective means of detecting an extraterrestrial civilization may be to search for random signals resulting from its technology. For example, thousands of transmitters on Earth radiate into space and the radiation may reach the vicinity of the stars nearest to us. The more advanced the civilization is, the more powerful may be the radiation accidentally transmitted in various directions.

One cannot exclude the existence of advanced civilizations which utilize energy and, consequently, radiate a considerable fraction of it, and exist sufficiently close to us—within 10 to 1000 light years from the Sun. Thus we can search for signs of the technology of an extraterrestrial civilization when making observations of irregular radio emissions from space at different wavelengths. Recognizing the slight probability of such success, we nevertheless believe that this work may lead to new astrophysical discoveries since sporadic radio emission also may be of natural origin; e.g. associated with explosive processes in our Galaxy, or with the occurrence of the annihilation (conversion into radiation) of matter and antimatter.

The question concerning the possible existence of radio pulses that accompany bursts of gravity waves escaping from the Galaxy (discovered by J. Weber late in the 1960's) has also been discussed in the literature. It would be very interesting to discover such pulses and, in 1970, English investigators made parallel observations at wavelengths from 2 to 7 m. However, they did not detect any pulses from space, nor any other new phenomena.

Unlike the English scientists, we believed it better to try to detect sporadic signals at centimeter and decimeter wavelengths, where atmospheric and industrial noises are much reduced. For example, the interference at decimeter and centimeter wavelengths caused by lightening at a height of about 2 km has an effective radius of nearly 150 km. As for industrial noises, the sources on Earth are noticeable only out to a distance of several kilometers. At wavelengths longer than tens of meters (and partly so at meter wavelengths), the situation is different. Interference signals are propagated throughout the spherical layer bounded by the ionosphere and the Earth's surface.

There are other, more concrete motives for a study of sporadic radio emissions at centimeter and decimeter wavelengths. The fact is that in radio astronomical observations with long integration time, we have often recorded bursts of radio emission and irregular, extended oscillations in the intensity of interference. Usually, it was caused by local industry (electric welding, automobile ignition, loose contacts in the power supply, switching of everyday electrical devices, etc.) or, in the case of especially long-time interferences, defects in the apparatus.

For radioastronomical observations, large directional antennas—radiotelescopes—were used. Pointing the radiotelescope at different regions of the sky or the ground, we attempted to find the direction from which the interference originated. It appeared that the noises came from practically everywhere.

Regarding interference at centimeter and decimeter wavelengths, there is a firm notion that they may be either of industrial Earth origin or caused by atmospheric electricity and sometimes also by auroras. Regardless of this idea, it was suggested that the interference registered in radio astronomical observations came from outer space. Thus the above considerations and facts led to the conclusion that it would be expedient to search for sporadic radio emissions from outer space.

Methods of search

To investigate the questions considered above, measurements were made at the Radiophysical Research Institute during 1969–70. Since the signals may be expected from any direction, and from the Earth's surface, reception must use a simple linear antenna capable of capturing radio emissions from practically all directions. Further, to separate deep space signals from local noises of technical origin and from atmospheric noise, reception must be carried out simultaneously at several sites far distant from each other. We chose Ussuriysk, Kara-Dag (the Crimea), Tuloma (Murmansk region) and Pustyn' (Gorki's region). It is obvious that the phenomenon of radiation coincident in time at various sites could not be caused by local sources if the number of events significantly exceeds the number of possible random coincidental events. Normal radiometers were used to receive the weak noise signals.

Searches for sporadic radio emissions were first carried out at a wavelength of 50 cm and then, after various interesting phenomena were detected, observations were made at wavelengths of 30, 16, 8 and 3 cm. The signals were recorded on paper tape recorders at about 10 mm/min. Time marks were made every 60 sec, enabling one to fix the simultaneity of events with an accuracy of 10 sec. The radiometers registered signals corresponding to effective sky radio temperature† of not less than 3° and not more than 300°K.

Simultaneous measurements were made from March to August 1970 at a wavelength of 50 cm at two sites: Kara-Dag and Pustyn' (radio astronomical stations of the Radiophysical Research Institute); from 1 September to 12 November of the same year, simultaneous observations were made at four sites—Pustyn', Kara-Dag, Tuloma and Ussuriysk—the measurements being made for 2 days, followed by an interval of 2 days. At the western sites, the antenna axes were inclined toward the east at 45° and, in the eastern sites, toward the west at the same angle so that the antennas were pointed at the same part of the sky.

In January–February 1971, measurements were made simultaneously at three stations in the Gorki region separated by 100 to 300 km. In June–October 1971, we carried out simultaneous observations at a series of wavelengths: 52, 28, 8 and 3 cm at Kara-Dag (the Crimea) and Zimenki (Gorki region). Finally, in July and August 1972, measurements were made at a wavelength of 50 cm. In this case, one of the reception sites was on the ship "Academician Kurchatov" cruising near the equator.

Observed phenomena and preliminary conclusions

The first measurements showed that there are signals observed simultaneously at various, even widely spaced, sites. All phenomena observed at each site may be divided into three types. The first type is single radio bursts of a duration of tens to hundreds of seconds, seen against the background of natural noises recorded by the receiving apparatus. The second type involves noise storms of different intensity and duration which represent chaotic oscillations of the signal intensity. The third type represents long changes in the mean radiation level.

Almost all observed phenomena were of the first type (bursts). Noise storms seldom occurred and the third type of variation was even rarer. We note that practically all the observed bursts and noise storms with intensities of 10°–300°K

†The burst intensity is characterized by the effective sky temperature, i.e. by the temperature of an absolutely black body in the form of the sky hemisphere which produces in the antenna the same power as the source of the burst. Since the bursts have a continuous spectrum and the source dimensions are unknown, it is essential that the received flux be expressed in terms of the equivalent radiation from the whole sky hemisphere, assigning to it an equivalent brightness temperature.

cannot be explained by sporadic solar radio emission. Otherwise the intensity of radio emission from the Sun must exceed its quiet level by 2–250 times. (At a wavelength of 50 cm, the radio emission from the quiet Sun could cause an antenna temperature of not more than 5°.)

According to the Solar Service data for the entire observation period in 1970–72, two comparatively intense radio bursts were observed on 6 September 1970 and on 4 August 1972. Both bursts were recorded in our observations.

The processing of observational results involved the determination of the statistical characteristics of bursts at each site, the search for phenomena occurring simultaneously at various sites, and the estimation of their statistical parameters. The examples of bursts as well as the character of their coincidences are shown in Figs. 1 and 2. Figure 3 illustrates the triple coincidence of a noise storm.

From 1 September 1970 to November 1971 (80 days of measurements), nearly 4500 bursts were registered at a wavelength of 50 cm and approximately 1000 bursts at other wavelengths. Noise storms of different duration occurred in 20 cases. Among all bursts at a wavelength of 50 cm, there were nearly 250 double and 130 triple coincidences. These data were used to estimate the statistical properties of bursts and noise storms, and their nature.

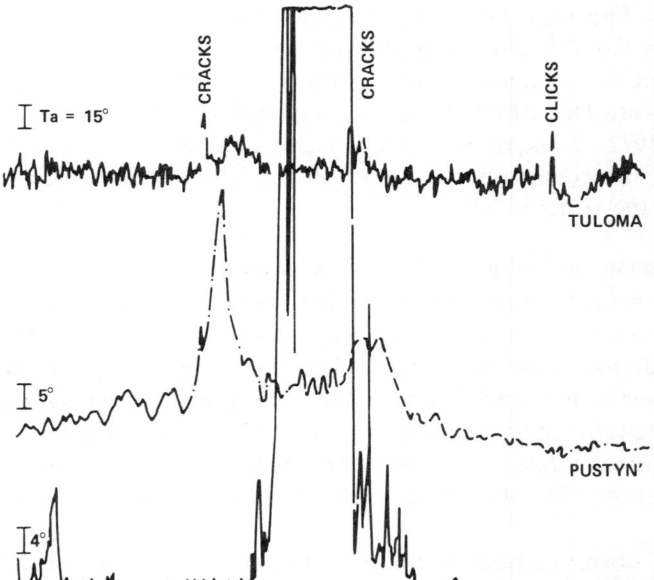

Fig. 1. Coincidence of bursts at 50 cm wavelength at three sites: Pustiyn', Ussuriysk and Tuloma (15 October 1970).

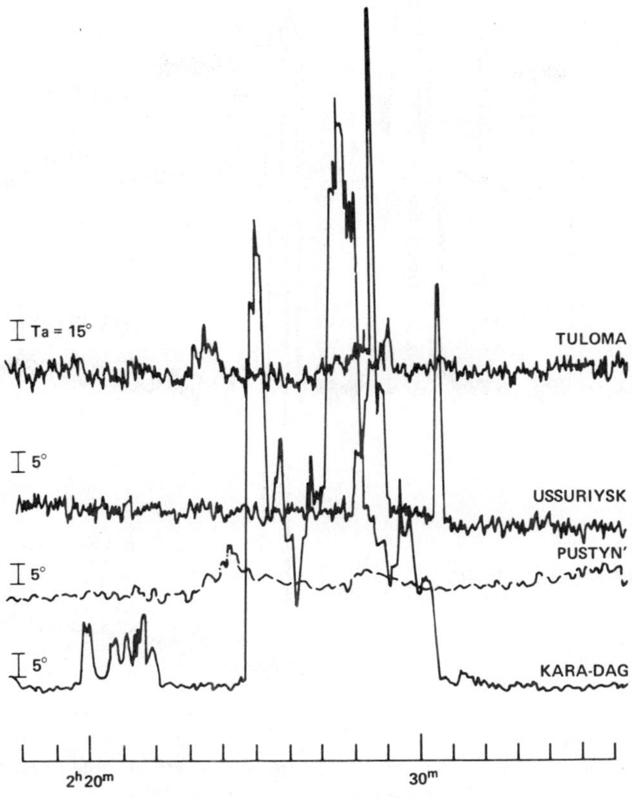

Fig. 2. Coincidence of bursts at 50 cm wavelength at four sites: Pustiyn', Kara-Dag, Ussuriysk and Tuloma (16 October 1970).

The basic conclusion is that the observed radio bursts and noise storms come from the visible hemisphere of circumterrestrial space rather than from far space. As a result, the forms of all observed coincident bursts are not the same. This means that the radiation source has considerable angular dimensions, i.e. it is not far from us. With bursts generated at distances much greater than the Earth's diameter and, moreover, at the distances of the nearest stars, the form of coincident bursts at all sites must be the same as it took place, for example, when the solar radio burst was received on 6 September 1970. Therefore, the assumption of the cosmic origin of bursts (if we mean here by the word "cosmic" the region beginning, at least, far beyond the magnetosphere of Earth) does not hold.

It became clear that the generally accepted hypothesis of the existence only of atmospheric and technological noises over this wavelength range is incorrect. The supposition was made that the observed radio bursts (at least those that coincide at various sites) are but the result of excitation of oscillations in the magnetosphere of the Earth and that the other bursts may be considered local industrial noises. Let us consider now in detail the main features of this newly discovered, sporadic radio emission.

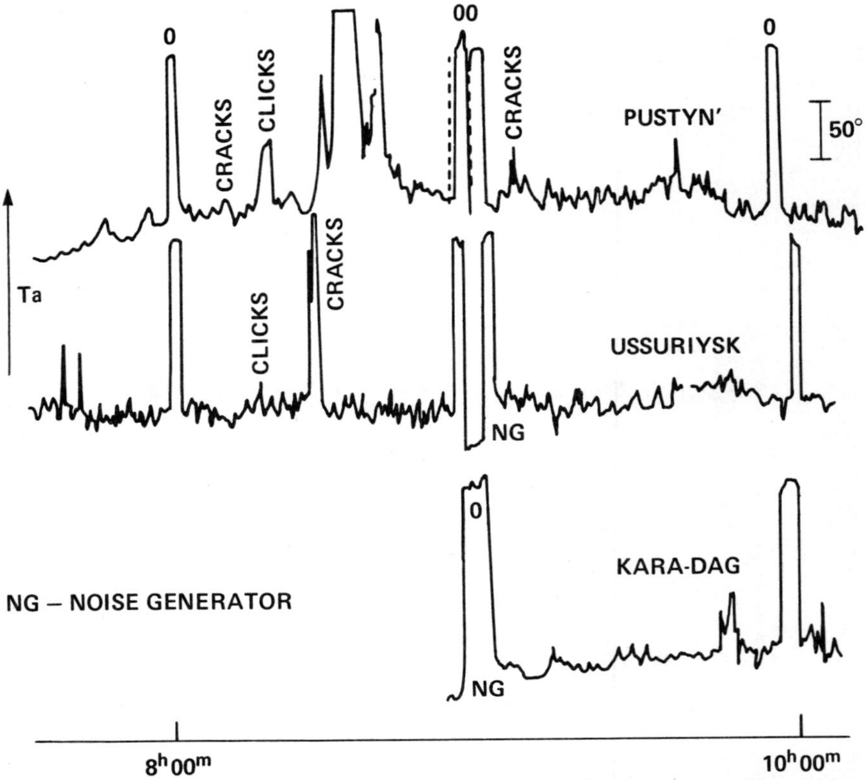

Fig. 3. Coincidence of the noise storm at 50 cm wavelength at three sites: Ussuriysk, Gorki and the Crimea (2 September 1970).

Basic experimental results

It appears that the duration and intensity of bursts are grouped in rather narrow ranges of values. Typical curves of the intensity and duration distribution of bursts are given in Fig. 4.

For a particular epoch of observations, the average duration of a burst remains practically equal at all sites and at all wavelengths used. The distribution law and average duration of a burst change essentially only from one epoch of observation to another. So, at all four stations in autumn 1970, the average duration of bursts at a wavelength of 50 cm appeared the same, amounting to 2–3 min. Observations made at the same wavelength in June and October 1971 in the Crimea and Gorki showed that the average duration was nearly 1 min.

The most frequently observed intensities of bursts at all wavelengths and sites are identical, from 40 to 50°K. Thus the radiation flux at 50 cm is approximately 8 to 10 times greater than that of the quiet Sun and several hundred times greater than the flux from the most powerful discrete radio source Cassiopeia-A at the same wavelength. The average values of burst intensities are also close to each other at various wavelengths and sites. We note a strong similarity between the intensity distribution curves at various sites at the same observing time, and the occurrence of a weak secondary maximum in the

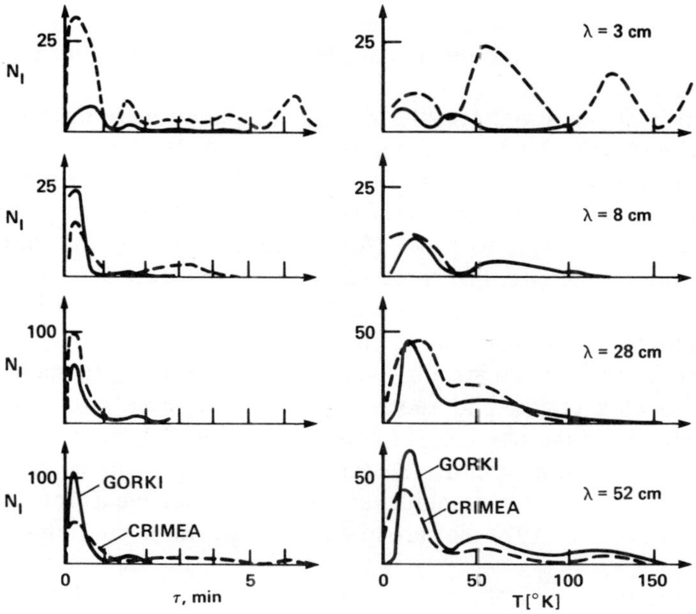

Fig. 4. Distribution over duration (τ) and intensity (T) of the total number of bursts (N_1) in Gorki and the Crimea for October 1971 at wavelengths of 3, 8, 28 and 52 cm.

distribution curve for observations during June and October 1971. We also observed variations in the duration distribution curves.

Of interest is the diurnal variation of the burst number distribution. It appears that the burst number density (number/hr) increases essentially in the daytime. The growth begins shortly before sunrise and a sharp fall is seen soon after sunset. Observations at 50 cm in Gorki and the Crimea showed that the daytime density was, on the average, 2 to 3 times greater than that at night. In September–November 1971, 3 bursts/hr were registered. For Ussuriysk and Murmansk, this increase was 3 to 4 times. The curve of the diurnal variation in Ussuriysk appeared to be shifted with respect to the curve in Gorki by the difference in the local time of these stations (nearly 7 hr).

At a shorter wavelength, the daytime maximum became less pronounced and, for example, in Kara-Dag was practically absent at 16 cm. In this case, one can speak about morning–evening maxima that are clearly seen at 50 cm, but less distinctly at 16 cm.

From the data obtained follows a conclusion quite unexpected from the point of view of the old concept of interferences at microwavelengths. The main characteristics of bursts for the same epoch of observations do not depend basically on the observational sites, regardless of a rather large difference in their geographical locations and the technological noise conditions. Thus, most of the observed bursts of sporadic radio emission are not the result of human engineering but are of natural origin.

Let us consider now the basic characteristics of bursts observed simultaneously at two or three stations at a single wavelength, or at one site at two or three wavelengths.

The bursts coinciding only partly in time (they may have different duration and intensity) are referred to as simultaneous (coincident) bursts. The percentage of double coincidences appeared to be rather constant over the whole year of observation and, in the first approximation, independent of wavelength, amounting, on the average, to 5–6% (for a pair of sites) of the number of all bursts registered at one site. For three stations, two independent pairs appeared at each site and the percentage of double coincidences of bursts will then amount to 10–12% (excluding triple coincidences).

A number of triple coincidences were observed for September–November 1970 at 50 cm wavelength, and those coincident at three sites were, on the average, nearly 3% of the number of bursts at one site. For three observing sites, the total number of double and triple coincidences at each site is about 14%. We observed also two quadruple coincidences for 10 days of measurements from 15 October to 6 November 1970 that are about 0.5% of the mean number of bursts per site.

The observations show that the mean burst duration and mean intensity (expressed in brightness temperature) do not depend distinctly on wavelength in the 50–3 cm range. A distinctly expressed dependence on wavelength is observed only for the mean diurnal burst number density. We observed that the burst number density increased in approximate proportion to wavelength.

The spectral width of the bursts is defined from a comparison of the number of coincidences at one site at different pairs of wavelengths: 50–35 cm, 52–28 cm and 52–8 cm. The fact that the largest number of coincidences accounts for the first pair of wavelengths and sharply drops at the following pairs testifies to the comparatively small width of the burst spectrum, evidently lying somewhere between 0.2 and 1.0 GHz.

The polarization of bursts was investigated by receiving with two orthogonal antennas situated in the directions north–south, east–west. Analysis showed that the burst radiation is essentially not polarized.

To determine the reasons for coincident bursts, one must determine whether these coincidences are real or accidental phenomena. For this purpose, one must estimate the number of accidental coincidences. With this aim, we compared the theoretical number of accidental double and triple coincidences with the number of coincident bursts actually observed.

The corresponding calculations for each two-day interval of observations for September–November 1970 at 50 cm wavelength show that the number of real double coincidences is always larger than the calculated accidental rate. The relation of these numbers varies from 1 to 3 over a given period of observation. For the same period, the relationship of the number of real triple coincidences for two days to the number of calculated accidental ones ranges from 3 to 10. The mean ratio for a two-day interval of the real number of coincidences to the accidental ones for the double coincidences is equal to 2 and, for the triple ones, is about 5. Thus, the coincidences observed are not accidental events and are generated by some global process of oscillation excitation.

To understand the nature of the global processes observed, it is interesting to investigate the diurnal variation of double and triple coincidences and its comparison with the calculated ones for the accidental coincidences. Figure 5 gives the mean diurnal distribution of the number density of doubly coincident events (number of coincidences in 1 hr) at Kara-Dag and Pustiyn' in comparison with the calculated number of expected random coincidences for September–November 1970. It is seen that there are two maxima of nonaccidental coincidences—in the morning and in the evening—where the excess is 5 to 7 times the accidental number of coincidences. At night, the number of double coincidences corresponds to the accidental rate.

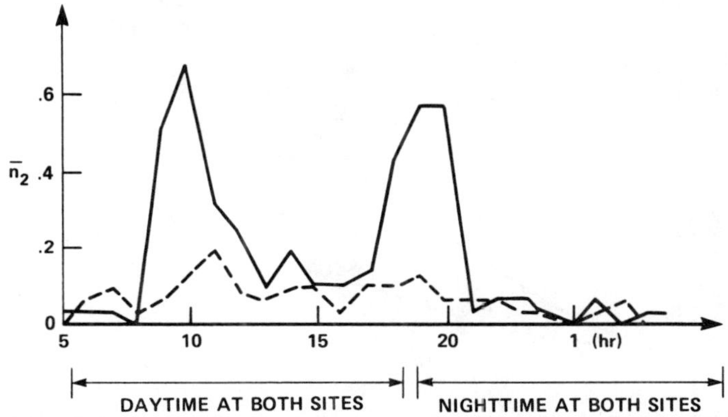

Fig. 5. Diurnal variation of the real and probable (random) number per hour of coincident bursts (density) $\bar{n}_2$ at 50 cm wavelength for Pustiyn' and Kara-Dag, averaged over the period from 1 September to 2 November 1970. Solid curve is the real number of coincidences; dashed curve is the calculated number of coincidences.

These data testify to the fact that the observed phenomena of sporadic radio emission have a deep physical nature. Evidently, we observed here the excitation of oscillations in the Earth's ion magnetosphere or in the upper atmosphere. The most probable reason for sporadic radio emissions may be its generation in the Earth's ion magnetosphere due to different corpuscular fluxes from the Sun and from the radiation belts. To verify this hypothesis, we compared in detail the characteristics of radio emission bursts with different occurrences of solar activity.

Sporadic radio emission and solar activity

To establish the relationship between the phenomena observed and solar activity, we compared the variation of burst characteristics during the entire observation period from March 1970 to October 1971 with solar activity data. This interval covers part of an 11-yr cycle when solar activity data changed noticeably.

We compared the time variation of monthly mean number density of bursts in 1 hr at either site with the time variation of monthly mean area of spots and the area of chromospheric flares on the Sun. As a result, correlations were obtained between different processes (Figs. 6 and 7). They are presented as the dependence of monthly mean number density of bursts on the monthly mean area of spots and chromospheric flares. A markedly different slope of the curves for Gorki and the Crimea may be explained by the latitude effect; however, it is possible that this fact is associated with partial covering of the horizon in the Crimea by the mountains in the north.

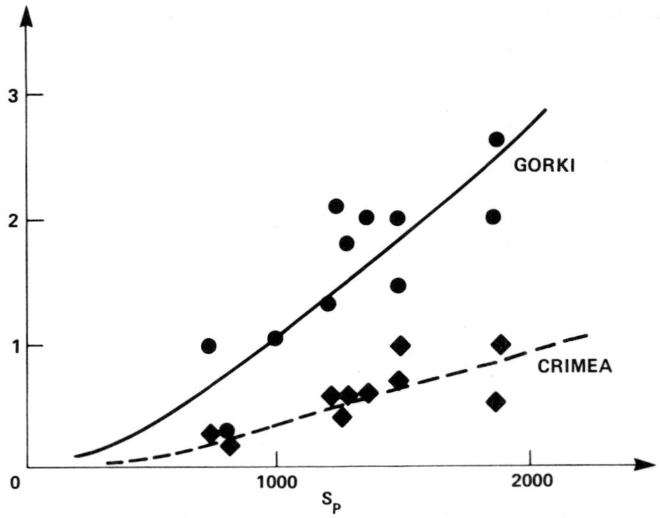

Fig. 6. Dependence of burst number density at 50 cm wavelength on the area of Sun spots for 1970–71; $\bar{n}_1$ = number of bursts per hour (averaged for the month) and S_p = area covered by spots, millionth of solar halfsphere (averaged for the month).

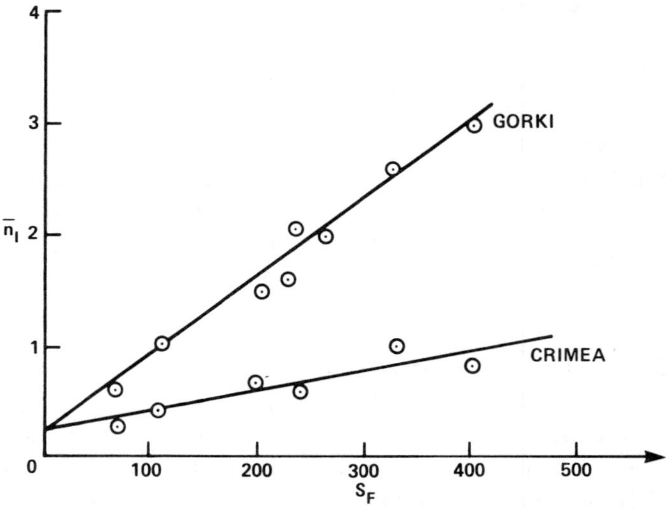

Fig. 7. Dependence of the burst number density at 50 cm wavelength on the area of solar chromospheric flares for 1970–71; n_1 = number of bursts per hour (averaged for the month), and S_F = area of chromospheric flares in square grad (averaged for the month).

Analogously, we found connections with solar activity. The monthly mean number density of coincident bursts showed an almost square dependence of density both on the spot area and on the chromospheric flares.

Thus we may say that, on the average, there is a close relationship between the number of bursts and solar activity. At the same time, it is interesting to make a more detailed comparison (with smaller averaging period).

We compared curves of the time dependence of the burst number for 2 days at 50 cm wavelength with the time dependence of the spot and flare area for September–November 1970 and found a similarity in these curves. For example, Fig. 8 compares the change of chromospheric flare area S_F with the total number

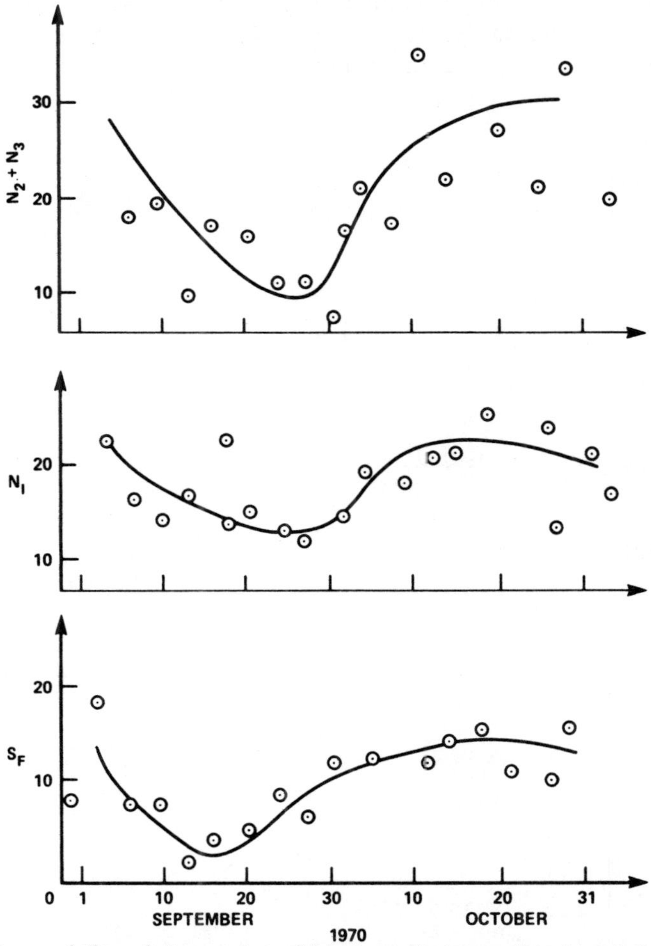

Fig. 8. Time variation of the number of bursts at 50 cm wavelength and the area of chromospheric flares for September–October 1970; N_1 = total number of bursts for 2 days of observations at three sites; N_2 = total number of double-burst coincidences (Pustiyn'–Kara-Dag, Pustiyn'–Ussuriysk and Ussuriysk–Kara-Dag) for 2 days of observations; N_3 = number of triple-burst coincidences at three sites for 2 days of observations; and S_F = area of chromospheric flares in square grad.

of bursts N_I at three sites for a 2-day interval of observation during September–November 1970, and also with the total number of double and triple coincidences $N_2 + N_3$ (interval and period are the same). Noise storms are taken into account. They are reduced to the equivalent number of bursts by dividing the storm duration into the main burst duration. Figure 8 shows a good correlation of both the total number and the number of coincident bursts with the chromospheric flare area. Figure 9 compares the number of triple coincidences of bursts with the area of bright chromospheric flares, and exhibits a more distinct relation between both phenomena.

Thus there also exists a certain relationship between the number of bursts of sporadic radio emission and solar activity in the case of more detailed comparison over the essentially small averaging interval. However, individual comparisons of bursts and noise storms with close-in-time events on the Sun (chromospheric flares, spot occurrences) were unsuccessful.

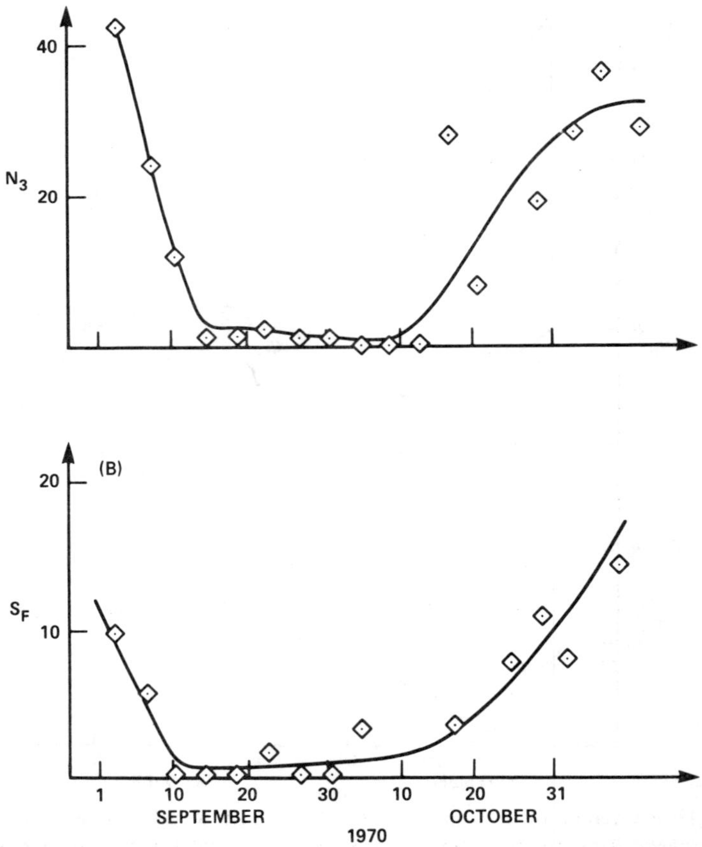

Fig. 9. Time variation of the number of triple-burst coincidences at three sites and the area of bright chromospheric flares for September–October 1970. N_3 = number of burst coincidences at three sites for 2 days of observations; $S_F^{(b)}$ = area of bright chromospheric flares in square grad.

Main conclusions: Nature of events

In summary, we may say that all our data, particularly the homogeneity of properties of radio emission bursts over a large territory and also the correlation of their occurrence with solar activity indices, show that the majority of radio emission bursts and noise storms observed at decimeter and centimeter wavelengths result from the sporadic radio emission generated in circumterrestrial space and not from industrial noise.

There are bursts of decimeter and centimeter radiation of global character that occur simultaneously at large distances. Global phenomena of sporadic radiation occur mainly during the daytime, most in the morning and evening hours. The existence of global bursts shows that generative processes may occur either at large heights on the order of some thousands of kilometers (the height of simultaneous visibility by Gorki and Ussuriysk) or at heights of tens and hundreds of kilometers when the region of generation covers large distances. The dissimilar character of the structural detail of bursts also testifies to the large scale and angular dimensions of the region responsible for radio emission.

Regarding the nature of the radiation discovered, we may make a number of assumptions. In principle, radio emission bursts may result from synchrotron radiation of relativistic electrons moving in the Earth's magnetosphere. Relativistic electrons are ejected by the Sun during chromospheric flares. In this case, the generation of bursts in the VHF band demands energies greater than 5–10 MeV. Yet, the experimentally proven value of the flux of relativistic electrons ejected by the Sun with energies greater than 5–10 MeV is 1/10 that required. Besides, the synchrotron radiation must be polarized and that was not observed.

Probably, we observe here bremsstrahlung radiation due to fluxes of charged particles braking in a sufficiently dense medium, e.g. in the upper atmosphere or in the lower part of the ionosphere. These particles may be electrons from the radiation belts in the process of particle precipitation in subpolar regions. Evidently, we may assume that observed radio emission bursts are due to short-period impulses of charged particles precipitated from the radiation belts in the lower layers of the ionosphere. It is well known that electron precipitation from the radiation belts accounts for the auroras. Also, the diurnal variation curve of the frequency of the auroral occurrences has maxima in the morning and evening hours similar to that shown in Fig. 5.

Thus, we evidently observe radio auroral flares with sufficient electron density taking place only at the moment of particle ejection. During the noise storm, the process of precipitation lasts about 10 min and sometimes for hours. However, a long-lasting noise storm is a rather rare phenomenon; e.g. during 2 months of observation (September–November 1970), only one storm lasted several hours. Note that, during observations near Murmansk at the time of the aurora, we registered a considerable increase in bursts at 50 cm wavelength against a background of a generally increasing sky brightness temperature up to 100°K.

It is known that particle precipitation from radiation belts occurs during a disturbance of the Earth's magnetic field structure. This process may be caused by solar particle fluxes with a "frozen" magnetic field. Such disturbances may be very short, creating the pulse particle ejection, or rather long, producing stable precipitation of particles in the upper layers of the atmosphere as a result of its magnetic storms and auroras. If this hypothesis about the mechanism of sporadic radiation is confirmed, then new possibilities will be opened for the investigation of geomagnetic phenomena, auroras, and processes in the Earth's ion magnetosphere. The investigation or radio emission bursts and noise storms permits a deeper penetration into the different aspects of solar-terrestrial phenomena.

Our investigations allow us to make a number of valuable conclusions on the character and methods of radiations from extraterrestrial civilizations.

It is rather attractive to conduct simultaneous observations at different sites on the Earth's surface using directional and nondirectional radiotelescopes. It is desirable to embrace the territory of the entire globe to provide a survey of the whole sky.

For nondirectional observations, the most suitable frequency band is the centimeter wave range where the noise level from the detected radio emission of the circumterrestrial medium is sufficiently small. To detect space signals during these observations, it is reasonable to use not only the features of simultaneity and similarity of the signal form at different sites, but also to use dispersion of the wave propagation velocity in space. The latter manifests itself in pulse time shifting at different frequencies. The distance from the radiating object may be defined by this shift. This permits one to separate with confidence signals from deep space from signals arising in the Earth's ion magnetosphere or in the solar system.

Finally, observations must be carried out with nondirectional antennas located on the Moon, on orbiting laboratories, or on interplanetary stations.

Evidently, the level of noises in space is essentially smaller than is found in observations from Earth. This permits one to increase essentially the sensitivity and reliability of the search for signals from extraterrestrial civilizations.

All these investigations are planned and will be carried out in the future, taking our experiences into account.

References

Shklovskii I. S. (1965) *Universe, Life, Intelligence.* Nauka, Moscow.

Troitskii V. S. *et al.* (1973) Search of sporadic radio emission from space at centimeter and decimeter wavelengths. *Izv. Vuzov. Radiofizika* **3**.

Acta Astronautica. Vol. 6, pp. 95–104. Pergamon Press 1979. Printed in Great Britain

Search for signals from extraterrestrial civilizations by the method of synchronous dispersion reception

L. M. GINDILIS

P. K. Shternberg State Astronomical Institute, Moscow State University, Moscow, U.S.S.R.

AND

N. S. KARDASHEV, V. A. SOGLASNOV,
E. E. SPANGENBERG AND V. S. ETKIN

USSR Institute of Space Research, Moscow, U.S.S.R.

AND

V. G. MIROVSKII

V. I. Lenin Moscow State Pedagogical Institute, Moscow, U.S.S.R.

Abstract—It is suggested that optimal call signals of extraterrestrial civilizations are rare pulse signals in the decimeter and centimeter bands. Problems related to the optimization of equipment for the search for such signals are considered. To detect remote radio emission sources, highest sensitivity can probably be realized using the method of synchronized reception for two (or more) widely spaced stations and by selecting among received signals those with characteristic properties related to dispersion in intersteller matter. Staff members of the Institute of Space Research jointly with the Shternberg State Astronomical Institute of Moscow State University, the Moscow Power Engineering Institute, and the V. I. Lenin Moscow State Pedagogical Institute, prepared suitable equipment for the 350–550-MHz and 13,700–15,200-MHz frequency bands and, in 1972, carried out a series of measurements on a 3000-km base. The results obtained from a consideration of the experimental data are presented.

1. Formulation of problem: anticipated character of signals

THE METHOD used to search for signals from extraterrestrial civilizations (ETC) depends to a considerable degree on the assumed power potential of the transmitting civilization (sender). The most propitious case for detection is continuous isotropic antenna transmission of broadband signals ($\Delta\nu \sim \nu$) and omnidirectional reception. Power in excess of 10^{30} W is required to ensure communication within the Galaxy ($R \sim 10^5$ light years). If such power is not available to the sender, he cannot or does not find it necessary to use it for SETI purposes, and he will utilize more economical transmission methods.

One version consists of successive irradiation of various sectors of the sky using highly directional antennas. In this case, at the receiving end, an ETC signal will have the character of a random radio emission burst. The probability of its occurrence (or its repetition frequency) and also its duration, which are *a priori* unknown, are determined by the sender's program.

The capabilities of this method are limited, especially under mutual search conditions, since the search time in direction increases with the directivity of the transmitting antennas. Another method of reducing power is to use coding methods that allow the transmission of information (with sufficiently high reliability) at small signal/noise ratios.

Many methods exist which realize such communication systems by using noiselike signals. However, most of these methods require a knowledge of the sender's code at the receiving end. Generally, this condition is not satisfied until contact is established. This difficulty can be overcome by selecting from the set of possible communication systems one or several of the simplest systems. Two simple communication systems satisfy the requirements enumerated above: (1) a system with a variable on-off time ratio, which utilizes rare short pulses whereby high peak power that ensures a prescribed reliability at a low mean transmitter power is achieved due to concentration of energy in a short pulse and (2) a system with a quasimonochromatic signal in which the same conditions are realized due to energy concentration in a narrow spectral band.

The use of the first system seems preferable since the utilization of narrow-band signals requires a frequency search at the receiving end of the communication line and consequently increases the indeterminacy of the system. For the receiving end, the use of quasi-monochromatic signals sweeping in frequency is equivalent to system (1) since, after it passes through interstellar matter, a short broadband pulse becomes a sweeping signal because of dispersion. In contrast, sweeping over the frequency in system (1) opens up further possibilities for increasing the signal/noise ratio at the receiver due to the dispersion effect when delays compensating the lag in interstellar matter are introduced during transmission.

We note that the transmission mode of rare short pulses is advantageous for the transmitting end in the sense that peak power pulses are generated relatively easily by means of any of the known burst processes. The search equipment is also simplified to the highest degree at the receiving end. Finally, it is important to note that call signals appear as strong noise at the transmitting and receiving ends. At the transmitting end, continuous, very-high-power transmission (even if it is monochromatic) constitutes noise for all other radio equipment. The situation is similar at the receiving end. Since the peak power of the call signals probably exceeds the power of normal ETC data transmissions by many orders of magnitude, the call signals at the input of receivers will accordingly exceed the sensitivity threshold and, consequently, constitute interference for an optimally designed data-receiving system. For rare pulse signals, radiation is completely absent in the intervals between them and, consequently, there is no interference. So the intervals between the pulses can be used for the principal informative transmission.

Thus, it appears probable that until two-way contact is established, the ETC signal has the character of short-lived radio emission, possibly consisting of a series of short pulses with a large on-off time ratio.

Detection of call signals arriving from an unknown direction requires continuous monitoring of the entire sky. In the long term, it will probably be necessary to design detection systems that use highly effective antennas. In the first stage, the sky can be continuously observed by semidirectional or omnidirectional antennas. Of course, in this case we can only expect the detection of signals that are sufficiently high-powered.

Preferably, the search operations should be of definite astrophysical benefit. In this regard, it is of interest that the search for ETC signals can be conducted concurrently with the detection of sporadic pulsed radio emissions arising during natural bursts, such as supernova explosions (Colgate and Noerdlinger, 1971; Colgate, 1972, 1975). These types of experiments are now being carried out by various investigators (Charman *et al.*, 1970; Dube *et al.*, 1973; Delaney *et al.*, 1974; Huguenin and Moore, 1974; Mongain *et al.*, 1974; Patrige *et al.*, 1976; Troitskii *et al.*, 1973, 1974).

2. Method of synchronous dispersion reception

Under real conditions in which short pulsed signals are received on the Earth's surface, the main problem is to develop methods for eliminating the many kinds of radio interference. Ground, aircraft, and satellite radio transmitters, lightning discharges, spark discharges from industrial installations, and radio outbursts from the Sun and from the Earth's ionosphere are known sources of such sporadic noise.

The use of diversity reception followed by correlation analysis is very efficient in reducing the effect of this noise. Separation of receiving sites by distances on the order of 1000 km makes it possible to eliminate almost all noise generated from the Earth's surface and in the lower layers of the atmosphere and ionosphere. Of course, each observing site must be chosen in quiet regions because, in correlator-type reception, the number of random coincidences of independent noise bursts throughout the observation periods is proportional to the noise level at each point. Such low-level noise conditions probably occur in remote mountains far from industrial regions. A plateau surrounded by a ring of mountains along the horizon provides good protection from terrestrial noise propagating by way of tropospheric scattering. Estimates have shown that the smallest number of sites required for continuous observation of the celestial sphere—with the condition that every astronomical object be visible simultaneously at least at two points, the angle at which the horizon is obscured being no greater than 10°—is eight. The sites should be located at the corners of a cube inscribed on the Earth's spheroid. The distance between neighboring points (edge of the cube) is $d = 2R/\sqrt{3} = 7360$ km. The "total sky" as a percentage of the entire half-sphere area for two points represents $\sim 60\%$ (without taking into account the angle at which the horizon is obscured). It is remarkable that, if an object is simultaneously visible at least at two points, every event is observable simultaneously at four points. The latter results in a many-fold decrease in the probability of a random coincidence.

However, the capabilities of such a terrestrial detection system may be limited considerably because of the increasingly more powerful transmitters that are being mounted on spacecraft, in particular, because of the prospects of direct television relays from satellites (in the optimal band used to search for ETC signals). Therefore, the possibility of setting up receivers in space at a great distance from Earth should be kept in mind. In particular, lunar screened zones can be used advantageously for reliable shielding from terrestrial noise.

The interstellar dispersion effect is a characteristic property of arriving pulses traveling over interstellar distances. The high-frequency components of the pulse arrive at the reception point much earlier than the low-frequency components. The magnitude of the low-frequency lag is proportional to the distance from the source, and it can serve as a good indicator of its origin outside the Solar System. This phenomenon has been thoroughly studied at many observatories during observations of radio emission pulses from pulsars. The receiving equipment for measuring the lag effect must consist of several radio channels recording the instants at which the pulse appears in each channel.

3. Equipment and observations in the 1972 expedition

Special, highly sensitive equipment was developed for the first trials of synchronous dispersion reception of space signals.

Decimeter system

The decimeter system consisted of two sets of identical receiving systems. The antenna was a double helix antenna with a 130° radiation pattern (half-power beamwidth). The antenna effective area was about 0.1 m^2.

The radio receiver (noise temperature about 300°K; 350–550-MHz passband) had at the output four narrow-band filters tuned to center frequencies of 371.5, 408, 458.5 and 535 MHz. The filters had 5-MHz bandwidths. The strength of the incident radiations was recorded simultaneously in the broadband ($\Delta f =$ 200 MHz) and in the narrow-band channels on a five-channel paper-tape recorder, together with 1-sec time signals. The recorded low-frequency band-width was 0.1 to 10 Hz.

The frequencies of the narrow-band channels were selected in a way which ensured that the lag between neighboring channels (due to interstellar dispersion) was the same. The magnitude of the lag was

$$t_i - t_{i+1} = 1.7 \times 10^{-21} \int_0^R n_e dl$$

where n_e is the average free electron density along the signal path and R is the distance to the source of the signals. For a mean value of $n_e \simeq 10^{-1}\,\text{cm}^{-3}$ and $R = 3000$ light years, $t_i - t_{i+1} = 0.5$ sec and the lag between the extreme channels (out of four narrow-band channels) is 1.5 sec.

The 5σ sensitivity of the receiver along the principal axis of the antenna was $10^{-21}\,\text{W/m}^2\,\text{Hz}$ for the broadband channel and $3 \times 10^{-21}\,\text{W/m}^2\,\text{Hz}$ for the narrow-band channels.

Centimeter system

The centimeter equipment (a single set) consisted of an antenna in the form of two broad-beam wave guide terminals connected alternately in phase and antiphase to the receiver. The resulting system was an interferometer with oscillating lobes. Such a system provides optimal sensitivity for the indication of point radiation sources. The mean interferometer lobe width was 12 to 15°. The half-power envelope width was about 120°. The effective area of the antenna system was 3 cm^2.

A modular radiometer with a 14.5 ± 0.75-GHz reception band was used as the receiver. A parametric amplifier was mounted at the input and gave a system noise temperature of 120°K. The output time constant was 0.1 sec. The data were plotted on the same recorder as in the decimeter channel (with the same time reference). The 5σ sensitivity along the principal axis of the antenna was 2×10^{-20} W/m^2 Hz.

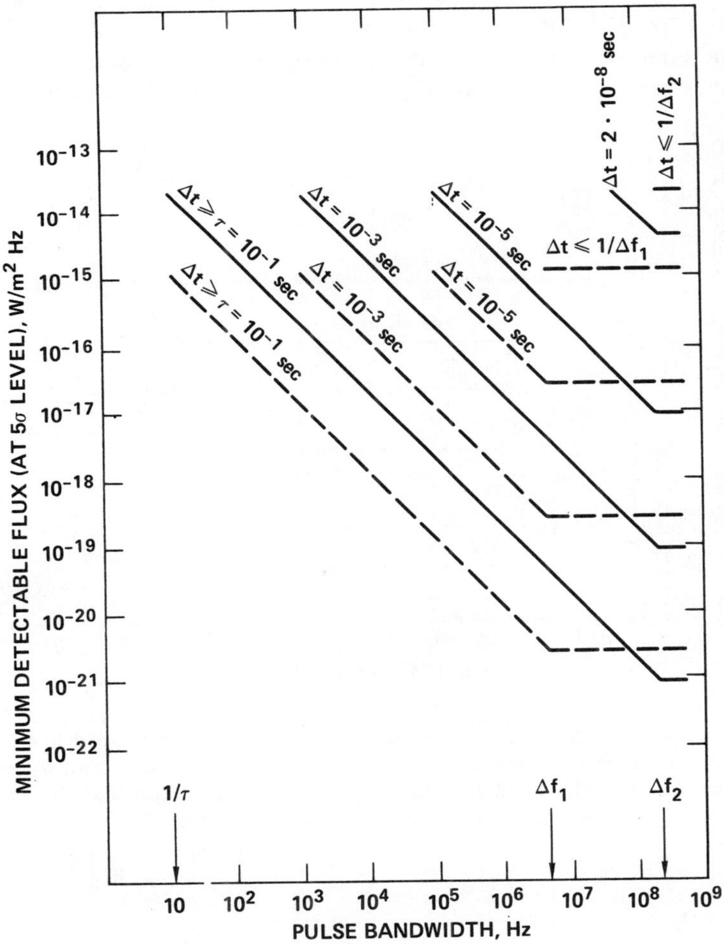

Fig. 1. Minimal detectable flux for decimeter receiver. Solid line (———) broadband channel, dotted line (-----) narrow-band channel. Arrows indicate bandwidth values Δf_1 for narrow-band channels and Δf_2, broad-band channels. The values of Δt and $\Delta \nu$ are given in seconds.

The data presented above characterize the sensitivity of the receiving equipment during the recording of a broadband signal which is continuous in time or of sufficiently long duration (greater than the time constant), occupying practically the entire range of the spectrum being investigated. For real signals, the detection conditions depend on the ratios of the signal parameters (duration Δt, frequency band $\Delta \nu$) on one hand and the reception equipment parameters (time constant τ, Δf band) on the other.

Figure 1 shows the smallest detectable flux F_ν (W/m^2 Hz) at the 5σ level for the decimeter receiver in the narrow-band and broadband channel for various pulse parameters. The minimum detectable pulse energy corresponding to this flux is shown in Fig. 2. The best detection conditions are realized for short ($1/\Delta \nu \leq \Delta t \leq \tau$) narrow-band ($\Delta \nu < \Delta f$) pulses. For these pulses, the minimal detectable energy is 1.5×10^{-15} J/m^2 in the narrow-band channel and 2×10^{-14} J/m^2 in the broadband channel. The corresponding energy values of the pulse at the input to the receiver are on the order of 1.5×10^{-16} and 2×10^{-15} J, respectively.

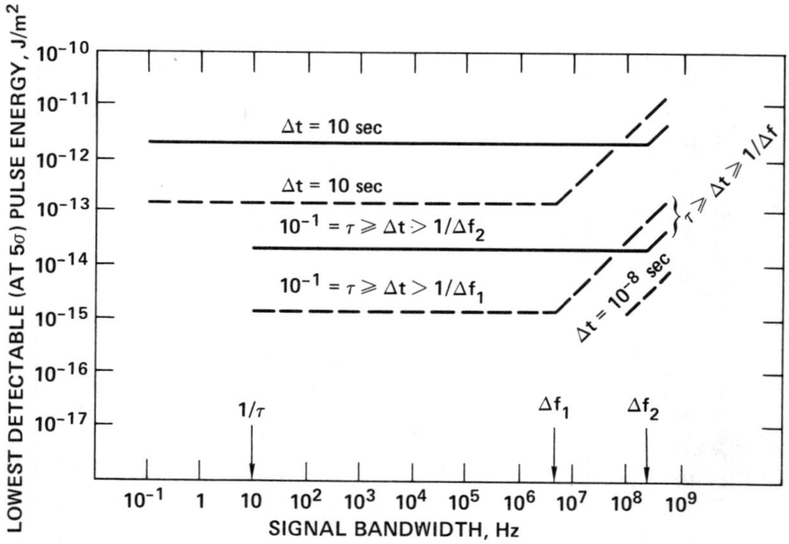

Fig. 2. Minimal detectable pulse energy for decimeter receiver. Solid line (———) broadband channel; dotted line (-----) narrow-band channel. Arrows indicate bandwidth values Δf_1 for narrow-band channels and Δf_2, broadband channels. The values of Δt and τ are given in seconds.

Observations

Two mountainous regions were selected for the observations. One site (latitude 43°35′, longitude 41°27′) is located in the Northern Caucasus Mountains in the Marukha River valley on the slope of the Pastukhov Mountain at an elevation of about 2000 m. The second site (latitude 37°18′, longitude 71°48′) is located in the Pamir Mountains in the Shardara River valley near the Roshtkal settlement, at an approximate elevation of 3000 m. The stations are about 3000 km apart.

Simultaneous 24-hr observations were made from 5 September to 25 October 1972. The net simultaneous observation time (after subtracting interruptions due to various causes) was 743 hr, or nearly 31 days.

Preliminary results†

A large number of diverse signals recorded by the decimeter channel was noted already during the observations. During its entire operating period, the centimeter channel recorded only several events, which (except for nearby lightning discharges) were uncorrelated with the record of the decimeter channel. Since there was only a single centimeter channel system, the possibility that these signals constituted nearby terrestrial noise cannot be excluded.

The simultaneous recording of signals in the form of rectangular pulses of 1.5-sec duration is shown in Fig. 3. These signals were observed simultaneously at both stations and over all channels. An attempt to measure dispersion for signals of this type has shown that the dispersion between extreme channels was less than 5 msec. A special high-speed recording experiment was performed for this purpose, in which a recording oscillograph was used. One can conclude that the distance to the radiation source is less than 10 light years. On the basis of many characteristics, it appears that the pulses are generated by transmitters mounted on artificial Earth satellites.

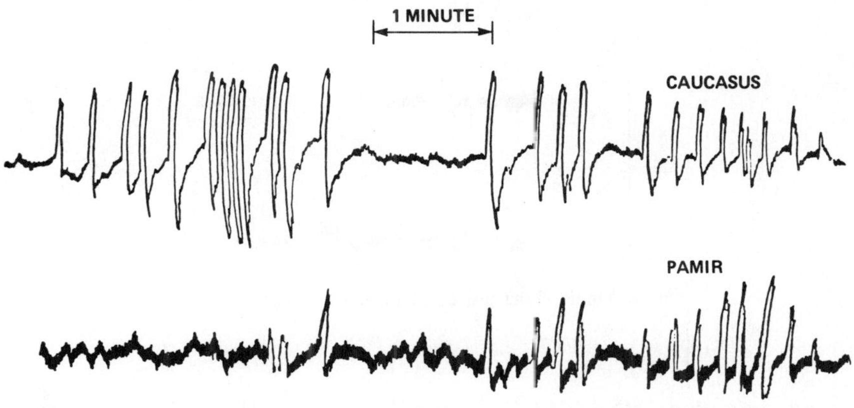

Fig. 3. Series of rectangular pulses coinciding at two points.

Figure 4 shows a second type of coincident signal, in the form of single short pulses. Throughout the entire observation period, only four cases were observed in which such signals coincided at two sites. Pulses that coincided were identified as solar outbursts. The remaining pulses of this type (single short pulses) were uncorrelated at the observation sites. They are probably related to local noise sources.

†For more details see Kardashev *et al.*, 1977.

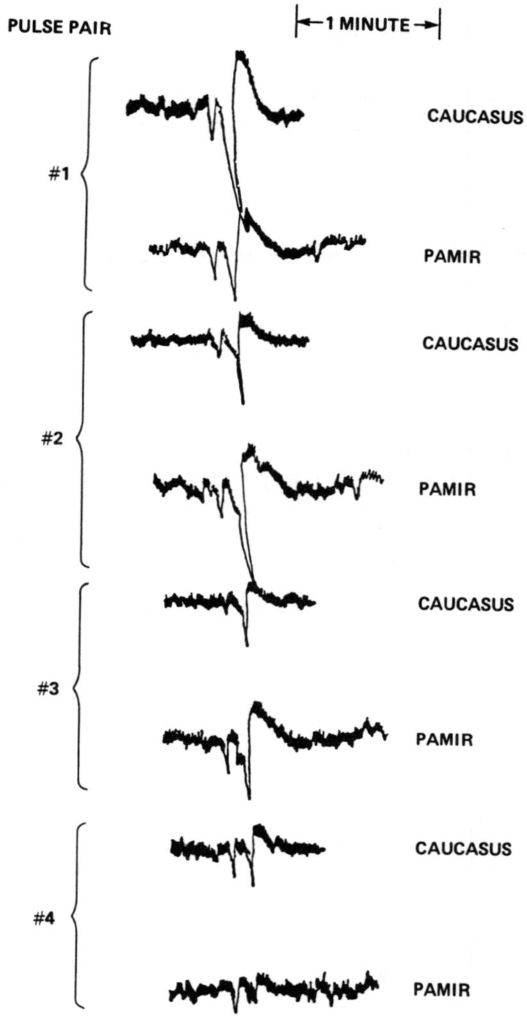

Fig. 4. Single short pulses identified as solar outbursts.

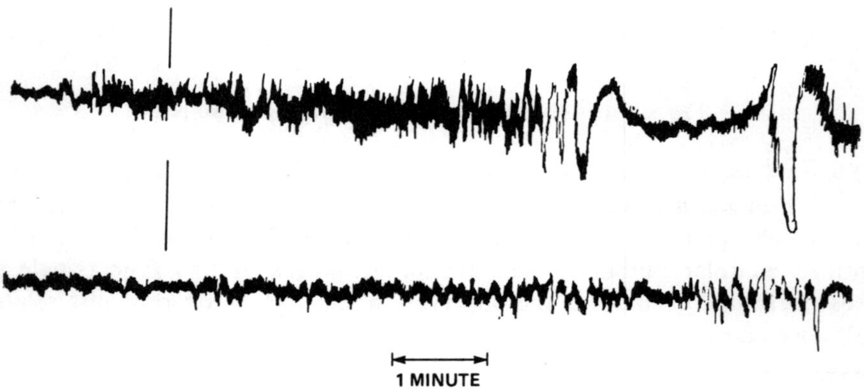

Fig. 5. Sporadic radio bursts, possibly of ionospheric origin (see Troitskii *et al.*, 1972).

Figure 5 shows a third type of recorded event whose shape resembles noise-like power fluctuations. Sometimes these signals coincided even in details. Sometimes they overlap only slightly in time and the shape of the signal is different at two points. In most cases, only a correlation in the mean was determined. A comparison with a similar study carried out at the N. L. Lobachevskii Scientific Research Institute of Radiophysics of Gor'kiy State University has shown that signals of this particular type apparently correspond to the sporadic ionospheric radiation detected there.

4. Conclusions

Not a single signal with marked dispersion which would allow one to assume that the source of the signal is located outside the Solar System was detected as a result of this experiment.

Further experiments with higher sensitivity and high-frequencies, capable of recording shorter pulses and suppressing terrestrial noise even further, must be set up in the future.

The following staff participated in the experiments:

USSR Institute of Space Research:

I. E. Val'ts	N. A. Savel'yeva
V. I. Vasik'kov	M. B. Semin
N. S. Kardashev	V. A. Soglasnov
N. Kh. Martirosyan	E. E. Spangenberg
T. A. Mizyakina	M. A. Strukov
Yu. S. Petrenko	V. N. Sysoyev
M. V. Popov	A. A. Tatarinov
A. S. Pshennikov	V. S. Etkin
I. G. Reznikov	

P. K. Shternberg State Astronomical Institute:
L. M. Gindilis
V. I. Konnikova

V. I. Lenin Moscow State Pedagogical Institute:
V. G. Mirovskii
V. V. Nikitin

Moscow Power Engineering Institute:
V. I. Yeremeyev

Acknowledgements—The authors thank Prof. I. S. Shklovskii for his constant, stimulating interest in this study and also A. Ye. Andriyevskii and V. N. P'yavchenko for their assistance during development of the equipment.

References

Colgate S. *et al.* (1972) *Ap. J. Lett.* **173**, L87.
Colgate S. and Noerdlinger P. (1971) *Ap. J.* **165**, 509.

Colgate S. (1975) *Ap. J.* **198**, 439.
Charman W. *et al.* (1970) *Nature* **228**, 346.
Delaney T. *et al.* (1974) *Astron. Astrophys.* **36**, 83.
Dube R. *et al.* (1973) *Nature Phys. Sci.* **245**, 17.
Huguenin G. and Moore E. (1974) *Ap. J. Lett.* **187**, L57.
Kardashev N. *et al.* (1977) *Astron. Zhurnal* (Russian Astron. Journ.) **54**, 3.
Mongain E. and Weeks T. (1974) *Publs. Astron. Soc. Pacific* **86**, 470.
Patrige R. *et al.* (1976) *Astron. Astrophys.* **46**, 477.
Troitskii V. *et al.* (1973) *Radiofizika* **16**, 3.
Troitskii V. *et al.* (1973) *Dokladi Akademii Nauk SSSR* **212**, 3.
Troitskii V. *et al.* (1974) *Uspekhi fizicheskikh nauk* **113**, 719.

Acta Astronautica. Vol. 6, pp. 105–127. Pergamon Press 1979. Printed in Great Britain

Strategic considerations in SETI, and a microwave approach†

CHARLES L. SEEGER

San Francisco State University, Department of Physics and Astronomy, San Francisco, CA 94132, U.S.A.

and

SETI Program Office, NASA Ames Research Center, Moffett Field, CA 94035, U.S.A.

Abstract—The only practical way we know now to actively test the hypothesis that life exists outside the Solar System, depends on an intelligent fraction of that life providing an electromagnetic signature we can recognize over interstellar distances. The nature of the local universe, a relatively mature microwave technology, recent digital solid state developments, and a minimal number of *ad hoc* assumptions, suggest a promising set of initial strategies of exploration for a range of possible electromagnetic artifacts. As a consequence, the Ames Research Center (ARC) and the Jet Propulsion Laboratory (JPL) are pursuing a moderately broad program using existing radiotelescopes and automated, state-of-the-art electronics, with the objective simply to detect and identify a relatively strong, relatively narrowband, essentially steady, intelligent, non-human, microwave signal. There are two main parts to this approach. In a companion paper, R. E. Edelson (JPL) discusses a survey over 80% of the sky in the 1–25 GHz band, with a frequency resolution of ~300 Hz and to a sensitivity in the range of 10^{-19} to 10^{-21} W/m^2. The second half of the approach (ARC) discussed here, is a study primarily of neighboring F,G,K, "target" stars in the 1.4–1.7 GHz band, with a frequency resolution of ~1 Hz and to a sensitivity of 10^{-25} to 10^{-27} W/m^2. Whether or not a detection is achieved, the combined approaches should reduce our present uncertainties by many orders of magnitude, as well as provide valuable astrophysical information.

I. Introduction

THE NOTION that intelligent social life exists elsewhere in the Galaxy and in the Universe, is a widely held hypothesis with now several decades of increasingly strong belief in its essential validity. But this is all an extrapolation based entirely on indirect evidence gathered to date from many disciplines by a single example of life, intelligent by self definition and ignorant of any physical laws prohibiting intelligent life elsewhere. To skeptics, Martin Rees retorted (Oliver and Billingham, 1972) "Absence of evidence is not evidence of absence". But of course, neither is it evidence of presence. Simply put, the situation is this. In spite of the many elaborate and purely anthropocentric discussions in the literature, there is no foreseeable way to test such a fundamental (but most appealing!) hypothesis other than to carry out dedicated explorations of the Universe beyond the Solar System. Even then, we should keep in mind that

†This paper, in slightly modified form, was presented at the XXVIII Congress of the I.A.F., Prague, 1977, as paper number 77–A–47.

though we use every bit of available intuition, science and technology in the search, the answer to our question may arrive serendipitously or by some extraterrestrial action totally beyond our intent and/or capability.

With deep interest in the validity of the hypothesis, the immediate central objective of a search for extraterrestrial intelligence (SETI) is the simple detection of a clearly non-human artifact. Already the possibilities for scientific exploration far outstrip our reasonably available resources. By and large, SETI efforts to date have used various equipments that lay at hand or were easily adapted to occasional SETI purposes, and it seems timely and appropriate to suggest a broad exploratory survey using state-of-the-art techniques at a level of effort promising orders of magnitude greater search sensitivity, but not at large cost.

It is a widely held view that the most practical and attractive way we know now to actively test the hypothesis that life exists outside the Solar System, depends on an intelligent fraction of that life providing an electromagnetic signature we can detect and recognize over interstellar distances (Oliver and Billingham, 1972). For this to occur, two conditions must be met:

(1) Their signature must arrive at Earth with sufficient intensity so that with our instruments and analytic capabilities, we can distinguish it from competing astrophysical and terrestrial radiations incident from the same apparent directions;

(2) We and they must possess in common certain minima of technical capacity, physical understanding of the Universe, and behavior patterns, both physical and intellectual.†

One may divide all possible electromagnetic signatures into two groups, incoherent and coherent. As examples of the former, there are the well known suggestions hypothesizing the thermal radiation from a Dyson civilization or other astroengineering activity, and the relatively narrowband but noise-like radiation from a controlled (modulated?) masering interstellar cloud. The coherent group is exemplified by the information probing and/or information bearing *signals* typical of our technology. At this stage of our technical culture one can hardly help being impressed by the spectral and power efficiencies permitted through the use of highly coherent radiations for signaling purposes.

There is but one "external" electromagnetic spectrum and its attractive uses are clearly manifold. Society has found the spectrum a singular, enormously valuable resource. Short of some sort of catastrophe, the odds seem strongly in favor of the Earth remaining electromagnetically bright at radio wavelengths—and for a very long time, even though the modes of spectrum usage can be expected to vary on a short time scale as technology and needs evolve. If we, why not some others, if others there be? Thus at the present, very elementary stage of human exploration of the Cosmos for ETI electromagnetic artifacts, and the while certainly not ignoring the incoherent (and other) possibilities, it does seem particularly appropriate to concentrate now on the detec-

†Of course, with trivial modification these necessary conditions hold for all possible communication media; and the details of the conditions will evolve as the societies concerned evolve.

tion of coherent signals which might be produced by another communicative technical society.

By design the survey outlined below is limited to the microwave electromagnetic spectrum. In this area our sensitivity is essentially limited by the nature of the physical universe, by economics, and by our imagination, but not by our technical proficiency; and it has been shown (Oliver and Billingham, 1972; Oliver, 1976; 1978) that within the limits of our understanding, information can be propagated successfully at these wavelengths over the greatest interstellar distances, assuming only a humanly manageable combination of transmitted energy, transmitting system, and receiving system.

Finally, though it is intrinsically not possible to say with much confidence, how and when, or even if, an extraterrestrial artifact will be (or is likely to be) detected *and* recognized, the studies described here are not open ended. The various facets of the survey search design block out clearly defined and significant volumes of multidimensional search space to be covered in a definite time period, after which a review is probably again in order. Additionally, there is appreciable anticipation of the potential astrophysical information which should accrue as a continuous by-product of this search for ETI, as well as technological spin-off's of potential value to several other human endeavours.

II. The situation

(a) The major dimensions of electromagnetic search space are signal strength, source directions, frequency, and modulation (i.e. any variation with time). The polarization dimension is easily encompassed. Observing in parallel at the six cardinal points of the Poincare sphere (Oliver and Billingham, 1972; Kraus, 1966), or serially for a continuous signal, the maximum signal loss is only 1 dB and the expected loss 0.39 dB.

(b) We know nothing about the characteristics of ETI microwave artifacts, nor how they are distributed across the spectrum. We assume (hypothesize) their existence, and in such form that we can detect and *recognize* them as distinct from intentional human radiations and from accidental or incidental radio frequency interference (RFI); and from astrophysical phenomena due to inanimate processes underway in the Universe. In fact and in principle, however, recognizing the artificial nature of a "signal" may be difficult. This is likely to be true for such hypothetical signals as the thermal radiation from a Dyson civilization. Such simplicity, or "naturalness", tends to let it be confused with other natural thermal sources. Signals of terrestrial or human origin should be simple to diagnose after a little observational experience, except for the intentionally fraudulent, for which SETI observatories should be forever on guard.† Artificial ET signals designed to convey appreciable information should contain a high degree of coherence, judging from our experience, though it is possible the coherence is in such a complex form that the total signal bears somewhat the characteristics of band limited thermal noise. In any case, the

†High resolution interferometric antenna arrangements obviously are useful in this connection.

range of signal possibilities just outlined should be kept in mind when designing SETI systems. For economy of effort, and because largely automated systems tend to sharply define the area in the modulation domain to which they respond in a desirable fashion, the signal analysis portion of a SETI receiver deserves the most sophisticated and thorough attention. And to date, this is the design area which appears to have been given the least detailed study.

(c) All possible ETI signals may be divided without prejudice into two classes, (i) those on which we might eavesdrop and, (ii) cooperative, or "beacon", signals intended to assist detection at great distances by parties unknown. In either case, assuming power detection and no prior knowledge about the signal format, the signals easiest to detect are those essentially constant in time and frequency which also provide high spectral power density.

(d) The nominal power sensitivity of a microwave receiver (or any "receiver" of electromagnetic energy) can be stated in the following way:

$$P_{\min} = kT_sB_r \qquad \text{(W)} \qquad (1)$$

Here, k is Boltzmann's constant (1.38^{-23} J/K); T_s is equivalent input system noise temperature in Kelvins (Oliver and Billingham, 1972; Oliver, 1976; Oliver, 1965); and B_r is the predetection equivalent power bandwidth of the receiver output channel concerned. Equation (1) assumes detection exists with a unity signal-to-noise ratio situation ($S/N = 1$), and that the signal stays within the bandwidth B_r for a time $\tau \geq \sim B_r^{-1}$ seconds, the channel response time. (For signals present in the channel for $\tau < \sim B_r^{-1}$ seconds, $P_{\min}$ increases in a predictable manner as τ decreases.) For any signal, maximum sensitivity occurs when B_r is adjusted to provide a maximum $\langle S/N \rangle$ when averaged over the channel response time. This value of B_r may well be less than the total transmitted signal bandwidth B_t. While present technology can achieve $B_r = B_t = 10^{-3}$ Hz, or less, in the 1–10 GHz range, suffice it to remark here that for the near future and for diverse practical reasons, one may venture the opinion that minimum receiver bandwidths are likely to be limited, give or take a factor of ten, to about one hertz; and ultra low-noise system technology appears to limit the maximum value of B_r to some hundreds of megahertz in the same frequency range.

(e) Closed cycle helium cryostat and maser technologies are sufficiently mature, reliable, and convenient in routine use, so that from the surface of the Earth, system noise is dominated by noise sources other than the input amplifier, over the 1–25 GHz range. For example, at 2.3 GHz a $T_s = 8.3$ K has been obtained by JPL with a low noise antenna pointed near the zenith (Claus and Wiebe, 1974). The maser was responsible for one-quarter of this temperature, the Galactic background radiation for one-half, and the remainder was due to atmospheric reradiation, antenna ohmic losses, and the surrounding terrain. Thus today, optimum T_s for frequencies as high as perhaps 10 GHz is essentially dominated by the nature of the Universe as observed in our neighborhood. As a result, system flux sensitivity

$$S_{\min} = P_{\min}A_e^{-1}(n\tau)^{-1/2} \qquad (\text{W/m}^2) \qquad (2)$$

is dominated by the antenna capture cross section A_e, and the integration time $n\tau$, for $n\tau \gg B_r^{-1}$, and not by our microwave and electronic technologies.

(f) Let $P = P_t G_t$ be the equivalent isotropic radiated power (EIRP) at the source of a distant signal, with P_t the real power emitted and G_t the directive gain of the transmitting antenna in the direction of the receiver, compared to the gain of an isotropic antenna. Then the EIRP required to produce a flux S at the Earth's surface from a distance of R light years (ly) is,

$$P = 1.12 \times 10^{33} SR^2 \qquad (W). \qquad (3)$$

A convenient way to visualize eqn (3) is this. One gigawatt at 10 ly distance will produce a flux on Earth of about 10^{-26} W/m^2. If the signal energy is evenly distributed across a one Hertz bandwidth, it will provide, in astronomical parlance, a flux density of one Jansky (Jy), or one flux unit (f.u.). In communications industry terms, $S = -260$ dBW/m^2. Assuming $T_s = 10$ K and $B_r = 1$ Hz, we see from eqns (1)–(3), that about 1.55×10^4 m^2 effective antenna area is required in order to provide unity signal-to-noise ratio in about one second of time at the detector of a SETI receiver. This area is about one-half that of the large antenna at the Arecibo Observatory in Puerto Rico. With a few tens of seconds integration time, this hypothetical signal should be easy to recognize against a white noise background.

(g) With respect to the direction of arrival dimension of the search, there are two dominent schools of thought. The classical approach originated with Cocconi and Morrison (1959; Cameron, 1963). It was refined and extended by the NASA Ames Research Center (ARC) study called Project Cyclops (Oliver and Billingham, 1972), in 1971, and has been further developed at ARC since 1974. This approach assumes that the most likely places to find microwave-detectable ETI societies are on planets orbiting neighboring more-or-less sun-like stars. It does not ignore other objects and directions, but merely urges giving major priority to systems likely to be similar to the Solar System, where we know an explosively developing, technically at least adequate, intelligent society exists, which has been advertising its presence effectively to its nearest neighbors for several decades! And it seems likely to continue to do so for perhaps a long time. Further, this approach recommends searching for both "leakage" signals and beacons.

The alternate approach prefers to make no assumptions about the sites of detectable ETI. The whole sky should be searched. Comparing these two approaches develops perhaps more subtleties than is immediately apparent at this point, particularly when frequency coverage is included in the discussion. (See below).

(h) We are concerned here with the terrestrial microwave window, about 0.8–25 GHz between the 3 dB points, as observed at a dry observing site. This is an enormous search dimension, from a technological point of view, because of the practical concern with frequency resolution in the range from a tenth Hertz to perhaps some hundreds of Hertz. With T_s limited by and large by properties of the Universe, flux sensitivity (S_{min}) is proportional to the ratio B_r/A_e. The resolution bandwidth may be limited by signal stability (e.g. Doppler drift and/or intrinsic variability); by multipath signal broadening and bandwidth limiting, both due to the turbulent interstellar plasma, effects known to be important for $R > 10^3$ ly in or near the Galactic plane, particularly at the lower microwave frequencies; and by the patience of the observer because, for example, recognizing a unity signal-to-noise ratio signal (see eqn (1)) with $B_r = 0.01$ Hz, takes about three hours, in order to achieve a low false alarm probability. Thus a maximum of perhaps 3×10^3 observations per year would be possible, and there are over 10^6 sun-like stars within 1000 ly of the Sun, and many more unique pointing directions in the sky when using a large antenna system. Therefore, without special information, B_r in practice is likely to be limited to about 1 Hz or greater. Hence, too, the instinctive, almost insidious wish for ever larger antenna systems!

(i) Maser technology with and/or without helium cooled up-converters can provide at least 300 MHz instantaneous bandwidth throughout the frequency range of interest, without serious degradation of T_s. Low loss (i.e. low noise), efficient antenna feeds are presently limited to about 20% bandwidths. At 1.5 GHz this allows a 300 MHz instantaneous receiver bandwidth which bridges the spectrum between the H and OH interstellar lines, 1420–1727 MHz—the "water hole" (Oliver, 1976; 1978). With such capability, it is fortunate that it is now practical to construct digital solid state, LSI, comb frequency filters with 10^6 to 10^{10} output channels, and to analyze these outputs in real time.

(j) Discussion of the choice of an "optimum" frequency band for initial SETI efforts has produced a spate of suggestions, often accompanied by tacit, as well as explicit, *ad hoc* assumptions about what "they" will "obviously do".† Again, however, there are dominant proposals.

First, there is the classic proposal to search in the direction of relatively nearby solar-type stars, with particularly high sensitivity for spectrally dense, coherent signals, and throughout a selected portion of the terrestrial microwave window. This was originally suggested by Cocconi and Morrison (1959; Cameron, 1963): "Therefore we think it most promising to search in the *neighborhood* of 1420 Mc/sec" (emphasis added). This has evolved through the Cyclops Study and through recent ARC analyses, to the Water Hole concept which urges major *initial* priority to searching for both class (i) and class (ii) signals in the 1.4–1.727 GHz band, and with the greatest practical sensitivity. This priority is also strongly supported by a SETI Science Workshop resolution included in the two year SETI Science Workshop final report (1978) recently forwarded to the Space Sciences Committee of the United States National Academy of Sciences. (In this connection see paragraph (k) below).

†Surveying the literature one may get the impression of an intuitive spree wherein arguments supporting this or that almost magical frequency band, are clothed in rational academic discourse as though the selected assumptions were known, high probability foundations worthy of major superstructure. It would be largely obvious nonsense to be discouraged were it not an historical fact that intuitive leaps have in the past proved to be highly productive. One may hope, though, for a more generous recognition of just how little we know of the capabilities and motivation of ET societies.

The basic assumptions here are that it may be likely that many societies in the Galaxy, once they have achieved our degree of technical competencies, may not only continue to evolve but will also continue as radio bright phenomena over long periods of time even though they exploit the electromagnetic spectrum with a sophistication as yet beyond us. Further, there is the typically (western?) human hope that some of these advanced societies may, for any of a number of conceivable or as yet unimagined motives, provide beacon signals at a modest power level equivalent, say, to 10^9 W radiated isotropically (which we could now achieve), which would intentionally, or inadvertently, assist our discovery of their existance if they were at distances from us as great as a thousand or more light years. (For further discussions, see Oliver and Billingham, 1972, Appendix B, pp. 180–181).

Second, there are the proposals to look for signals from super civilizations committing astroengineering on a large scale and radiating enormous powers, stellar or galactic in magnitude. Dyson (1966) suggested looking at infrared wavelengths for objects radiating stellar energies in the form of ~ 300 K thermal radiation. Kardashev (1971; 1979) and others, concentrating more or less on the entire electromagnetic spectrum, suggest closely studying (for signs of intelligence) all "anomalously powerful" or "suspicious" objects radiating enormous energy, particularly at infrared and radio wavelengths. Their discussions include monochromatic radiations, but strongly emphasize very broadband, continuous and pulsed radiations. The breadth and reach of their observational proposals are unusual because in a very real sense they are bold recommendations for orders of magnitude increase in the effort devoted to the human examinations of the Universe by electromagnetic means. This would accelerate an expansion in the objectives of classical astrophysics already under way; viz. to include more studies relevant to the origin and existence of life in the Universe.

Third, there is a compromise proposal; explore all directions in space over the entire microwave window, for powerful, frequency stable, moderately narrow band, continuous signals which though detectable with present equipment, could have been missed just because no one had searched for them. As Oliver (1978) has shown, to be of reasonable duration such a survey must sacrifice sensitivity for coverage in direction and frequency. Thus there is implied (or stated) an assumption that the signal source is very nearby, or extraordinarly powerful and so must be produced by a supercivilization; or "they" know where we are (and perhaps our stage of development) and are beaming a signal at us. Such a frequency survey would automatically acquire enormous information on the current human usage of the microwave spectrum, particularly with regard to satellites. With such a survey, it is not clear just how much of the window is occupied by SETI-blinding transmissions, but it is certainly a large fraction, and one which is expanding rapidly.

Since we truly know nothing directly about the matter of ETI, how one chooses among these approaches and their variants, is largely an intuitive, practical matter in which one's total background and psychological point of view is paramount. What does seem certain is this. The wider and more thorough the explorations, the sooner will there be pertinent data at hand. Certainly at the present time, and considering the paucity of our total observations of the Universe in the scant few years we have been at it, to argue forcefully that there is no one out there because we have not detected them and they have not visited us, does appear to be a leap into the wild blue yonder on a par with its opposite extreme. There does not seem to be any recourse open to humankind other than to carry out an exploration on as many fronts as possible. And it does seem desirable to do so now, from practical, scientific, and metaphysical points of view.

(k) By historical accident here on Earth, the 1.4–1.727 GHz Water Hole band is presently largely free of terrestrial transmissions which would seriously impair SETI studies in this band. A careful review of a relatively complete listing of U.S. licensed and cooperative installations revealed only a few ground based transmitters in the 10–100 W range and a few thousands in the $1–10^3$ mW domain, plus an uncertain contingent of avionic gear. In fact, spurious emissions from powerful transmissions outside the band may be the most annoying sources of interference. In addition, there are a few satellite down links of small bandwidth, and a few daily radiosondes. All told, investment and activity in this band is modest, perhaps negligible, due to normal obsolescence, and if minor protection can be afforded an intrinsically quiet SETI site.

Satellite down transmission, however, are another matter. The minor present satellite use of this band can be allowed to die out naturally in the next five to ten years. Imperative for SETI over the next few decades at least, is no additional use of this band by satellite systems now in the proposal stage or already in the experimental pipeline. *If allowed to continue, preparations now underway for massive use of this band by expensive multiple satellite systems is most likely to prohibit any significant SETI observations in this band from the surface of the Earth for the foreseeable future.*

For this reason the CCIR (Consultative Committee for International Radio of the International Telecommunications Union (ITU), an organ of the United Nations) has accepted for study a question on the use of a band for SETI. To achieve protection over the next several decades requires forceful approval by a majority of the national representatives at the next (1979) World Administrative Radio Conference (WARC) organized by the ITU, where it is one nation, one vote. SETI in the Water Hole band is possible now, but by the 1980s will become nearly impossible if no positive action is taken by the WARC in 1979, at least until technology and resources are sufficient to begin work again at a station in *distant* earth orbit. (See Appendix I).

SETI cannot use effectively the small, widely distributed radio astronomy bands, but these bands are mute evidence of trenchant history. Over the years since the 1959 WARC, astronomers have gained ever increasing and useful protection for the radio astronomy frequency bands, which are greater in percentage of the available spectrum than that desired for SETI. Thus, a twenty year precedent demonstrates that it is possible for an international organization and its constituent nations to show effective respect for the intellectual needs of human kind. WARC will respond to its varied constituency if the latter makes its reasonable wishes clearly known. Thus it is to be hoped—should be expected—that this respect will be demonstrated once again in 1979.

(l) The preceding notes have outlined some salient factors affecting the nature of the strategy underlying the five year search survey designs outlined in the next section, but some accidental bounds have not been mentioned. In particular, a serious observational design, more often than not, must defer from the pristine ideal in favor of estimated optimum results when confronted with the realities of available personnel and physical resources, funding prospects, and the need to interweave smoothly with other programs in and around common premises. This is normal and often important, as is suitable adjustment to the ambient human influences effective in the area of concern; and it is likewise seldom directly recognized in formal discourse.

III. The survey designs

Several outstanding facts dominate the microwave survey designs discussed here.

(1) No searches of neighboring star systems to date, have been able to achieve the sensitivities possible in the last few years, not by a factor of 30 dB and more. The required equipment has not been available.

(2) Astronomical whole sky surveys have been few, and only a minute fraction of the 1–25 GHz band has been sampled. Telescope time is in short supply and data handling facilities often awkward, seldom operating completely in real time. Coherent interference is a routine hazard, so when it is recognized the records are cleaned or destroyed. Observers tend to notice what they are looking for and what they are accustomed to expect. Without a special awareness, ETI signals would not be distinguished from terrestrial interference.

(3) Not even the Water Hole has been surveyed for (improbably?) strong ETI signals. Considering the 1–25 GHz range, I would hazard the guess that it is possible, even perhaps probable, that if a strong and steady, slowly modulated ETI signal has been present at a level perhaps as high as $-150\ \mathrm{dBW/m^2}$, it would have gone undetected though it had been around for years.

(4) Prof. Alvin Despain (University of California at Berkeley) and Prof. Allen Peterson (Stanford University) and his colleagues, have demonstrated that digital technology has reached the point where on-line spectral analysis and signal discrimination is not only practical but economically superior to rival technologies when considering frequency filters on the order of megachannels and up.

(5) When making great jumps in observational technology, it is wise to acquire observational experience as early in a building program as may be possible. This is particularly true for SETI, where the effort eventually required is quite uncertain. Thus it is accepted that no new antenna systems will be suggested at this time. Rather, existing installations will be used in order to look for strong, and in the main simple, ETI signals, and to gain experience with the dramatically new electronic hardware.

(6) The modulation/time search domain was briefly discussed in the previous Section II–b. For the eavesdropping mode, if carriers are not present, the

possibilities are almost dismaying. With carriers present, both modes tend to merge. In any case, we know what we would look for at first—high spectral density, coherent, essentially extremely stable, point source signals with perhaps some accompanying sidebands. That these are the easiest to detect and identify, hence first to be searched for, has been recognized for almost two decades (Cocconi and Morrison, 1959; Oliver and Billingham, 1972; Oliver, 1976; 1978). For what it is worth, we note that Earth is most easily detectable at the distance of the nearest stars through its UHF TV carriers.

(7) The possible reality of ETI beacons has demonstrated attraction, as a literature survey will demonstrate. The Cyclops Report (1972) and B. M. Oliver (1976; 1978) have shown that, interstellar plasma multipath effects aside, the general spectral region including the Water Hole is an obvious choice for beacons as a result of purely physical arguments and one assumption, viz. that the ETI society also recognized the value of capital, and operational, economy of effort (as we visualize it). Out to a sphere of radius 10^3 ly, which encompasses $> 10^6$ sun-like stars, the plasma multipath phenomenon is probably insignificant in most if not all directions. Further narrowing the optimum spectral region to the band between the interstellar H and OH lines is a practical matter of the present moment (for technological, economic, and terrestrial spectrum usage reasons) as well as suggesting a second assumption. "They" may also, as do some of us, appreciate the cosmic sign post aspect of these prominent emission lines (Oliver and Billingham, 1972; Oliver, 1976; 1978).

(8) These instrumental and search survey designs have been developed over the past several years in close cooperation with scientists at JPL and ARC, and the NASA coloration is intentional since it is hoped these two research centers will be able to carry these designs to fruition in the near future, perhaps in the period 1979–84. Whether NASA should move into this area is a decision not yet firm.

(9) Table 1 outlines the major features of the two search system designs. Design No. 1 is not discussed in detail here, being covered (at an early stage of development) by R. E. Edelson (JPL), elsewhere in this volume. Both receiving systems are in line with good radio astronomical practice. It is planned that in each system, astrophysical data at the detector output will be compacted systematically and made available to all interested parties, along with the circumstances of all observing runs whether or not impaired or wiped out by faulty operation or RFI. By and large, JPL would be responsible for the wide-frequency all-sky survey, while ARC would concentrate mainly on Water Hole studies of nearby sunlike stars. But this would not be a sharp division. Considerable overlap is foreseen along with close cooperation in many areas. Finally, since both these approaches to SETI are still in the developmental stage, albeit fairly well along, technological advances and institutional decision may well change some aspects of the designs before an operational stage is reached. Thus the following overview of the ARC search system has somewhat the nature of a progress report; i.e. where we are in the development of an observational design for one aspect of SETI, an approach using existing radiotelescopes and automated, state-of-the-art electronics with the objective simply to detect and

Table 1. Summary of search designs

ITEMS	DESIGN No. 1 (JPL)	DESIGN No. 2 (ARC)
Main Search Modes	Frequency and Whole Sky	Targeted, Nearby Stars
Nominal Frequency Range	1–25 GHz	1.4–1.75 GHz
Instantaneous Bandwidth	≤ 300 MHz	≤ 10 MHz
Resolution Bandwidth (B_r)	≤ 300 Hz for 10^6 bins	≤ 4 Hz for 10^6 bins
Polarization Channels	2	2
Polarization Range	All Polarizations	All Polarizations
Output Channels (bins)	$\leq 10^6 \times 2$	$\leq 10^6 \times 2$
Nominal Sensitivity, W/m^2	$10^{-19} - 10^{-21}$, (all sky) 10^{-24} (targeted)	10^{-27} (targeted) 10^{-26} (all sky)
Time Integration Following Detection and Before Data Analysis	Tens of Seconds	B_r^{-1}
On-Line Data Handling	Almost All	Almost All
Preservation of Astrophysical Data	Yes	Yes
Signal Pattern Analysis Mainly On-Line	Yes	Yes
Visual and Interactive Pattern Analysis	N.A.	Available, *Not* Routine
Cryogenic Input Amplifier	Dual Up-Converters and Masers	Dual Tunable Masers
Antennas	JPL Horns and 26 m and 9 m Mirrors	Various Large Collectors
Sites	Goldstone, CA, USA	Arecibo and Others
Sky Coverage	80%	100%

identify a relatively strong, relatively narrowband, essentially steady, intelligent non-human microwave signal.

IV. The ARC search system design

The Ames SETI Program Office would propose to develop three complete receiving systems, outlined in Fig. 1, and keep them for long periods of time at a number of radio-quiet radio astronomy observatories with large antennas, e.g. the Arecibo Observatory of the National Astronomy and Ionospheric Center (NAIC), and the Ohio State-Ohio Wesleyan Radio Observatory (where SETI observations have been under way for some three years). The SETI program would share these facilities in the usual, cooperative fashion. In addition, when not in use for purely SETI purposes, the equipment in whole or in part, would be available for use in a variety of astrophysical programs of interest to astronomers also using the same observatories. When not thus in use, the signal-analytic portion of the receiver could ride piggyback on whatever receiver was in use at the time, and so achieve a double benefit. *First*, it would provide an extraordinarily sensitive means for detecting coherent RFI at low but significant levels that are hard to pinpoint in the usual radio astronomy system. (Some observatories have already indicated strong interest in this capability). *Second*, it would be able to carry on the SETI at somewhat reduced sensitivity, probably, but in whatever band was in use. Being wholly automatic in operation (except under alarm conditions), the ARC equipment should present minimal additional complications to the observing staff. One ARC objective is to disturb normal astrophysical research at the observatories just as little as can be arranged (indeed, to help wherever it can), while searching the regions around several hundred of the nearest sun-like stars (distributed over the whole sky) for artificial ET signals, mainly in the Water Hole of course, but whenever possible at other frequencies as well.

The block diagram of Fig. 2 shows the organization of the gain-stabilized receiver proper. Not shown are the diagnostic devices such as phase-lock-loop detectors, etc. required to examine suspect signals. Note the absence of input switching. To conserve observing time, precisely stable, calibrated gain, phase and frequency performance is achieved through solid construction, temperature

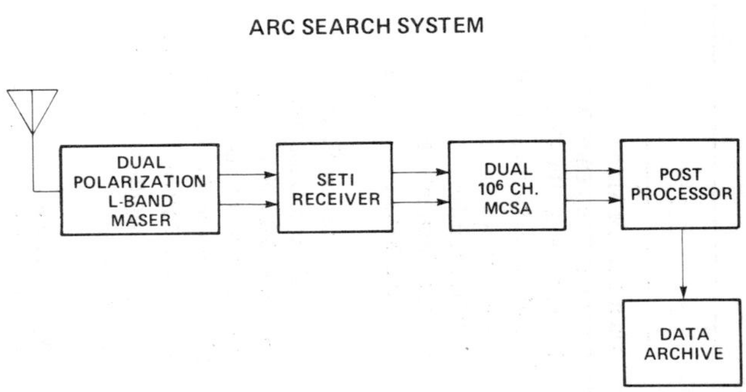

Fig. 1. Block diagram of the high sensitivity search system proposed by the SETI Program Office at the NASA Ames Research Center (ARC).

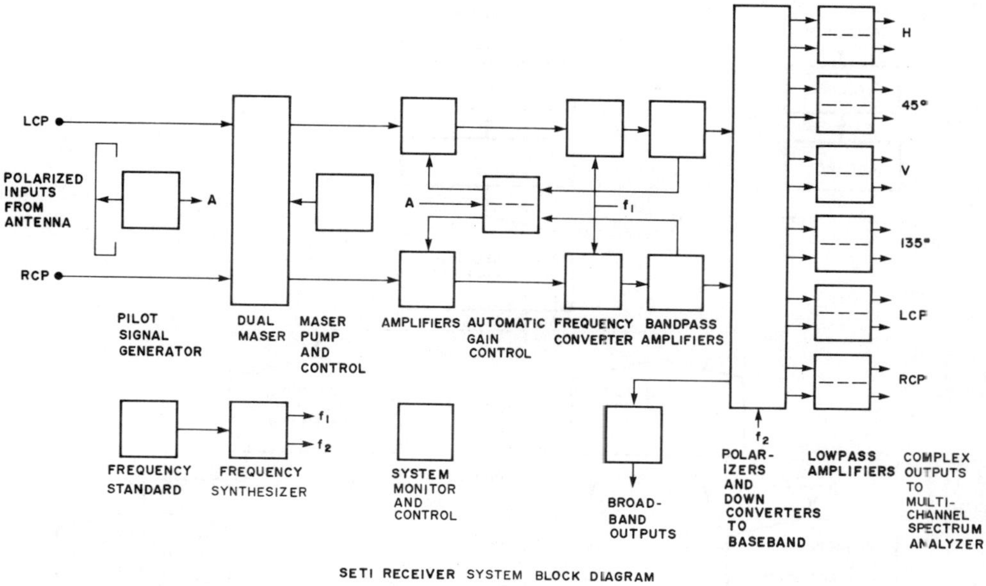

Fig. 2. Block diagram of ARC SETI receiver.

control, and a pulsed "standard" noise source adding only one to two percent additional system noise. The automatic gain control (AGC) system is designed to control long term drifts in performance, not short term instabilities which should be eliminated by other means.

Figure 3 outlines just one of the current solutions of providing 2^{20} output channels at low cost; and Fig. 4 shows some details of one of the parallel FFT output units. This architecture has several advantages in addition to the low cost made possible by the use of off-the-shelf chips already in mass production.

(1) There is a large degree of parallel operation, leading to simple, parallel control and increased reliability.

(2) The digital low-pass and bandpass filters greatly reduce the likelihood of aliasing by strong signals and further improve reliability, since many typical faults will affect only one-16th or one-128th of the operation. The technique used can provide linear phase-shift comb filters with steep skirts crossing at, say, the 3 dB points, and with >60 dB rejection of signals present in all other channels.

(3) The entire system can be under any desired degree of soft-ware and/or firm-ware (plug-in) control. The output binwidth can be controlled by either the clock rate or by a change in the number of FFT operations, within certain limits. For the particular design shown, chip choice sets a maximum input bandwidth of 4–5 MHz, thus 4–5 Hz output binwidth maximum at 2^{20} bins. But FFT commands can alter the output binwidths as long as (binwidth $\times$ number of bins) $\leq$ 4–5 MHz. Thus for this particular design, binwidth is limited to a maximum value of $(5 \times 8192) = 40{,}960$ Hz. Since the entire system (see Fig. 2) is under atomic

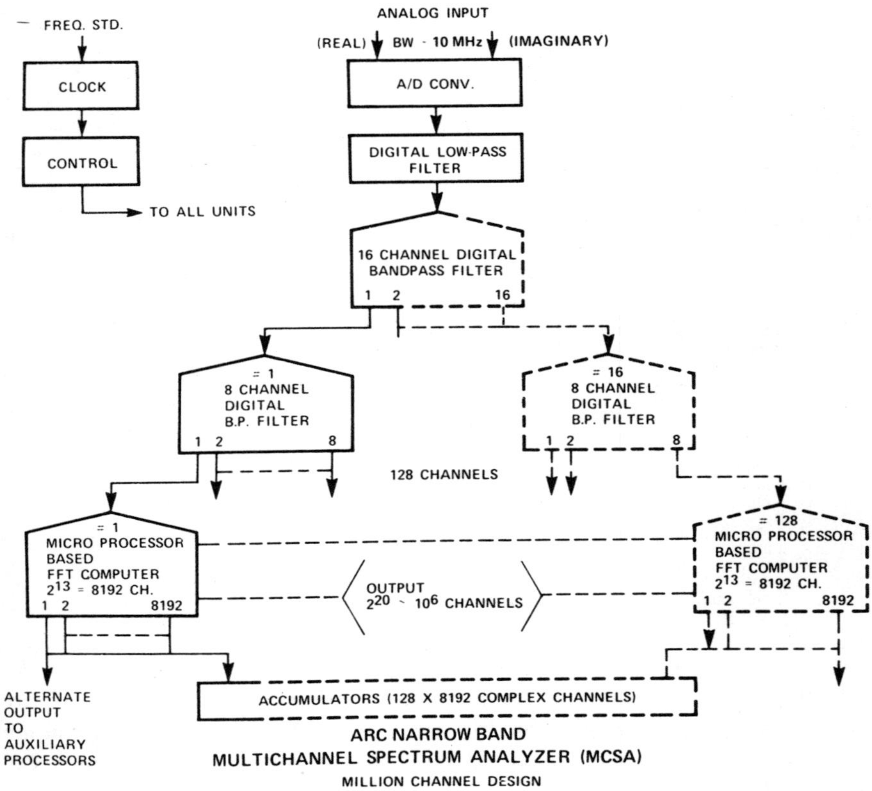

Fig. 3. Block diagram of ARC narrowband million channel spectrum analyzer (MCSA).

frequency-standard control, the minimum output binwidth can be made as small as 10^{-4} Hz.

As shown in Fig. 3, the output channels are complex (amplitude and phase). Detection has not yet occurred. When not operating at maximum input bandwidth, this design allows additional internal computation; and the output may be arranged in a variety of forms; e.g. $(a+jb)$; $(a^2+b^2)^{1/2}$ arctan (a/b); $(a^2+b^2)-$(a constant); $(1/n)\,\Sigma\,(a^2+b^2)i$, $(1 \le i \le n)$, where i can be a weighted function of frequency or time.

The major cost in this multichannel spectrum analyzer (MCSA) design lies in the memory required. This is a technological area which, as is well known, is still in a logarithmically decreasing cost/bit mode. Consequently, optimum design choice still depends on just when a design must be fixed so that final construction can get under way. New architecture and more economical and more efficient circuit designs are being developed more or less continuously by Prof. Allen M. Peterson and his colleagues and students at Stanford University, who should be contacted directly for further information on this attractive technology. At the present moment it appears likely that by the advent of 1980, the parts cost for a 2^{20} channel, one hertz binwidth MCSA may be less than \$30 K, a

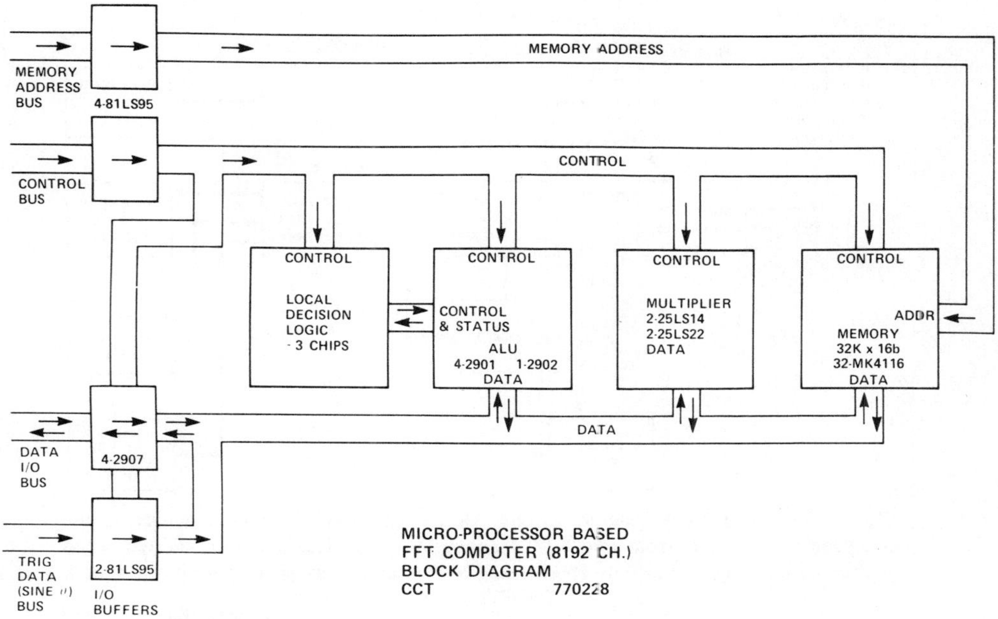

Fig. 4. Block diagram of a single 8192 channel microprocessor FFT computer board. 128 of these plug-in boards provide 2^{20} output channels in the MCSA of Fig. 3.

delightful improvement over the cumbersome technology necessarily chosen at the time of the 1971 Cyclops Project (Oliver and Billingham, 1972). Until such time as a design must be fixed, we will continue to explore the possibilities presented by new LSI chips, new algorithmic developments, and computer simulations of complex multistage dedicated processors. As a final note, this technology would seem to be attractive for similar solutions of this and other transforms (dispersion compensation, etc.) in other research and processing areas where large volumes of data must be treated essentially in real time.

The final element in the ARC design is the postprocessor, or signal extractor outlined in Fig. 5. It has not as yet received such detailed study as has the MCSA. The task here is to apply a variety of signal recognition and compaction algorithms to the flood of input bits from the MCSA's. Each MCSA bin produces a 32-bit complex work (16-bit + j 16-bit) each B_r^{-1} seconds. Besides a naturally conservative approach at this stage in the prototype system design (before observational experience), there are two reasons for the size of the MCSA output words.

(1) To preserve the ultra low noise property of the receiver and thus obtain an optimum S/N ratio, 4-bit digitizing is used at the input to the MCSA, along with stepped control of the average input signal in order to preserve approximately optimum digitizing levels in spite of a wide ranging average input signal level.

(2) The S/N ratio deteriorates in the FFT stages of the MCSA if too much round-off error is permitted. Further, if there is a strong signal in one or more

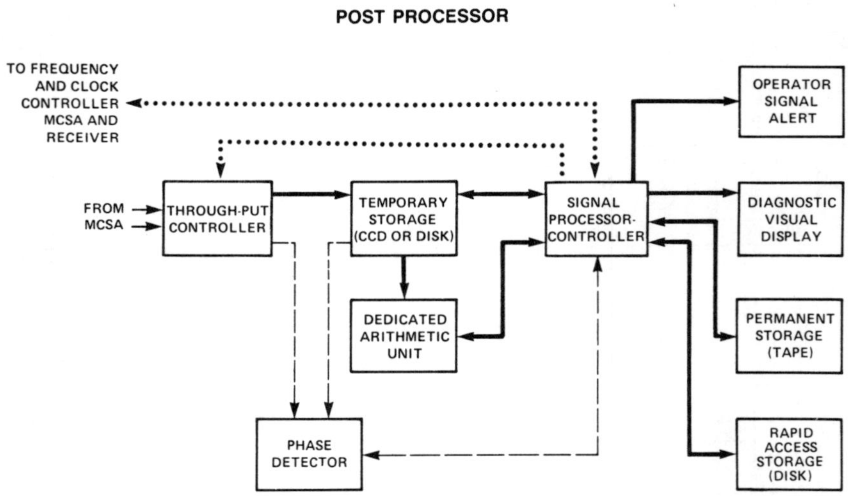

Fig. 5. Simplified block diagram of the ARC postprocessor, the automatic, on-line analyzer of the data from two 20^{20} channel MCSA's. The operator, alerted when malfunction or unusual signals obtain, may also interactively participate in the search for ETI signals.

channels, false signals can appear in otherwise signal-free channels. For the time being, we have a design which keeps this aliasing below a significant level even if the signal in one bin is perhaps 50–60 dB above noise. At least because of satellite down links in the Water Hole, we shall need this ability.

Not all the MCSA output bits need be used necessarily in the postprocessor. Since we are interested chiefly in small signal analysis, an occasional (rare) clipping of peak signals presents a negligible problem. Thus we can estimate roughly that for many operations 8-bits of output is sufficient, but 4-bits will often add undesirable noise. (Four-bit-slice architecture is common today in economical LSI components). Even after dropping half the MCSA bits, the postprocessor memory requirements exceed those of the MCSA by a large factor, at least for operations more complex than simple accumulation and peak recognition. Memory costs already are approaching \$100/M-bit in LSI CCD (large scale integrated, charge coupled devices), and further significant reductions are expected in the next several years. Thus as we shall see, technology is rapidly solving a major postprocessor cost problem.

The organization of the postprocessor is a compromise between cost and the range of signal types we would like to be able to recognize. In addition, because of the lack of experience, a large degree of flexibility is desired in these first experiments in automated signal recognition. The cost driver in the hardware design is temporary memory cost/bit. The receiver and MCSA provide 2^{21} output signals ($2^{20} \times 2$ polarizations) each averaged over a time $\simeq B_r^{-1}$ seconds (adjustable over a wide range, as noted above). In the first instance, since the most probable output of any channel is noise of almost vanishingly small information value, it is necessary to process this output for useful information worth saving for a long time, and to dump the remainder. Such a decision process is necessary and, of course, necessarily discards some information which might be found of value as a result of further study at a later time. To save the raw data of a few years of MCSA output is impractical in the extreme: To save raw data for up to some tens of minutes is attractive and possible.

Since a "signal", *ipso facto*, will require special study, which takes time, one must establish a satisfactorily low false alarm rate so that the search may progress at a satisfactory pace. The sources of false alarms should be two-fold, RFI and statistical peaks in the background band-limited white noise. Two approaches minimize the RFI problem. First, choose a quiet site. Second, in order to avoid local electromangetic interference (EMI) at the observatory, shield the receiving system with great care. More than the usual radio astronomy procedures are required because of the extreme sensitivity to stable, narrow-band, coherent signals.

With respect to statistical peaks in the receiver background noise, see B. M. Oliver's extensive discussion in the Cyclops Report (Oliver and Billingham, 1972). A single "unit observation", one recorded in a time $\simeq B_r^{-1}$ sec and present in one or more MCSA output channels, is essentially indeterminate in character by itself unless of sufficient strength, e.g. $S/N > 25$, and coupled with *a priori* knowledge. It is for this reason that careful observers integrating for a time $T \gg B_r^{-1}$ in order to increase sensitivity, always record data at intervals, Δt, such that $B_r^{-1} \leq \Delta t \ll T$. Only by examining the statistics of the unknown signal being accumulated, can one form a degree of reliable opinion about its character. Finally, to make full use of the data available at the output of the MCSA, the signals should be stored every B_r^{-1} seconds. Any form of averaging before analysis destroys information, narrowing the range of signals which may be recognized.

Figure 5 is a block diagram of the postprocessor, or signal extractor. The key element here is the temporary memory field, one half of which is analyzed while the other half is being filled. The dimensions of the temporary storage are frequency and time. Each B_r^{-1} seconds the MCSA's pass on 2^{21} data words in the form x_f/θ, filling a line ($t = \text{const.}$) with complex numbers organized by frequency and polarization. After 25–100 lines have been filled, it is possible to analyze the array for real signals in any MCSA channel, and with a false alarm probability of $10^{-9} - 10^{-12}$ if the signal-to-noise ratio at the detection is slightly greater than one.

The analysis of such a spectrum-time memory array is discussed in the Cyclops Report (Oliver and Billingham, 1972). The dedicated arithmetic unit time-averages the energies in each frequency band and calculates dispersions. It then repeats the process for grouped channels, where adjacent channels are combined in successive powers of two. By adding appropriate frequency offsets to the array, one then combs the field for narrow, drifting signals. The signal processor and controller (SPAC) then scans this data and decides on the proper course of action; routine compaction of the astronomical data; recognition of previously identified RFI; tripping of an alarm if routine examination shows unusual energy in a few channels (continuous or occasional, i.e. pulsed);

polarization calculation; passing data to the permanent storage magnetic tape(s) and clearing the memory fields if no alarm has been set; searching for drifting signals within suitable limits, and so forth. The SPAC will require only the power of a minicomputer if the bulk of the averaging and dispersion calculations are carried out in the special processor.

The calculations described so far use only one-half the input data, the amplitude (or power) in each MCSA channel. If the phase information in each channel can be used, coherent signals can be recognized at lower signal-to-noise ratios. (In any event, recognition of phase coherence in a possible signal will help with RFI identification.) T. Clark (NASA-Goddard Space Flight Center) brought this to our attention at Ames, B. M. Oliver calculated the phase statistics of a coherent signal as a function of S/N, and A. M. Despain (University of California at Berkeley) demonstrated in principle how digital phase-lock detectors could be used to scan the data field. No detailed designs have yet been developed, but it is clearly a promising area for further study.

The human being has exceptional sensitivity to patterns, so a flexible visual display is considered essential. It will be used at least for diagnostic purposes, to study both system operations and portions of the temporary memory field brought to the operator's attention by the automatic alert criteria. One property of the display unit is worth noting here. It can examine the temporary memory field with varying resolution. The frequency-time data in a thousand channels can be displayed, or adjacent channels may be summed to any desired degree and displayed, and coherence and polarization can be indicated. Through the use of an on-line, "instant replay," magnetic tape recorder, interesting but mystifying data may be saved for later, more sophisticated analysis.

Also available to the operator are tunable audio receivers, standard spectrum analyzers, phase-lock receivers, and the like, for direct study of the receiver outputs.

Since the entire system is operating at all times at sensitivities useful in radio astronomy, base levels, spectral data, and any time varying data will be stored in the permanent memory. While the entire system is designed specifically for SETI, it is believed this extra astronomical effort can be achieved at negligible additional cost.

Besides three complete systems to be used at various radio astronomy observatories, the prototype MCSA and postprocessor would be kept at Ames where they can be used to rapidly reduce digital tapes of data recorded at many places, not only by NASA scientists but by others who may find such a facility advantageous.

V. Conclusion

The theses advanced here may be summarized as follows. SETI is already recognized, and not in the United States alone, to be an important and appealing scientific exploration of inestimable value to all humanity. It presents some aspects of hunting for a needle in a hypergeometric haystack, but systematic search has found such needles before. Up to the present, SETI efforts have been left to a few dedicated astronomers without especially suitable equipment. Their persistence should be acknowledged. But technology has advanced rapidly in recent years and it is now eminently reasonable to suggest a modest program which would in a few hours outdo by orders of magnitude many of the past searches. More than that, it would solidly block out a significant volume of search space in a period of about five years, providing a solid foundation for reconsideration of the problem and perhaps the design of further explorations if no ETI signals have been detected.

SETI represents an expansion of the astrophysical investigation of the Universe and it is appropriate now to apply the full capabilities of modern technology to the problem, rather than to put it off many decades into the future when suitable space or moon stations may possibly be established. To pursue SETI now from the surface of the Earth and along the lines suggested above, represents a dual approach. (1) Survey the terrestrial microwave window with modest sensitivity and in all directions, for extraordinarily powerful transmissions which just possibly might be there and, also, be free of terrestrial interference, (2) simultaneously, but in the Water Hole, carry out a high sensitivity, necessarily targeted search (in the main), for weaker signals which may originate from the vicinity of nearby, sun-like stars. We know nothing about spectrum usage by ETI societies, so it is common sense to concentrate attention on that part of the spectrum where intrinsic sensitivity is the highest for the easiest signals to detect. After all, weaker signals are more probable than stronger ones. And despite attractive flights of imagination, the only solid evidence now available supports the notion of studying with maximum care, the vicinity of more-or-less sun-like stars. Searching for super civilizations (Dyson, 1966; Kardashev, 1971; Kardashev, 1979) may well be a very rewarding pursuit, particularly as outlined by Kardashev and his colleagues; but it must be noted that so far, the possibility and probability of super civilizations has in no way been proved or disproved, scientifically, technically, or sociologically. So it is perhaps well to carry out the exploration on many fronts.†

It is necessary to state here a sense of great urgency with respect to SETI. The demand for spectrum space by satellites and other services antithetical to SETI is enormous, and techniques for spectrum sharing between earth-to-earth and space-to-earth services, are developing so rapidly that one may look forward to a relatively near future when satellites transmit throughout most of the microwave window. As a consequence,

(1) Such search designs as discussed here should be carried out just as

†It is sometimes argued no advanced ETI societies exist, else they would have made themselves known to us, or we would have discovered them, before now. Therefore, SETI is doomed to failure and a waste of effort. Of all the presently imaginable possibilities when we are in such a state of near total ignorance, why argue that such an extremum is most probable? The blunt facts seem to be just these. Besides keeping our eyes open to the Universe, so to speak, only systematic exploration seems likely to significantly narrow down the near infinite possibilities, perhaps even develop some certainties. And in any event, experience has demonstrated the value to humans of exploring our environment to the limit of our ability. Space is now our frontier, and it would appear merely provident to explore it, perhaps even exploit some of it in good time.

quickly as can be managed; spectrum useful to SETI is being lost almost monthly;

(2) A window to the Universe, optimum for SETI, must be well supported at the 1979 World Administrative Radio Conference in Geneva.

Without such actions, a serious SETI is likely to be delayed indefinitely, or until Earth develops extensive, large engineering activities in space. This would seem to be an avoidable delay in human progress.

No suggestion about costs has been provided, so it remains to assure (only) that these "test bed" designs are well below the levels space agencies around the World accept as normal. There are the further expectations that such developments would be of value not only to astronomy, but to instrumentation and procedures in related fields. With respect to astronomy, useful astrophysical data is certain to be a by-product of SETI. The Water Hole contains H, OH, recombination hydrogen, and other interstellar lines, and perhaps yet unknown spectral features. (It has never been surveyed at the proposed sensitivity.) The ARC receiving system should provide significant advantages when used by radio astronomers between SETI runs, because of its extreme sensitivity, flexible spectral resolution, polarization discrimination, and built-in data-crunching abilities. No receiving system of comparable power exists today.

Acknowledgements—It remains to acknowledge my happy debt to all those with whom I have worked in the SETI area since the time of Project Cyclops. But it must suffice to recognize here just some highlights in the participation which has led to the present state of affairs. For major contributions, then, thanks to Bernard M. Oliver, John Billingham, Bruce Murray, Sebastian von Hoerner, Tom Clark, and the members of the ARC SETI Program Office and the JPL SETI Project Office; and professors Allen M. Peterson and Alvin M. Despain. Lastly, but intimately involved in the past two years with all of us, Philip Morrison and the members of the (SETI) Science Workshop, including Vera Buescher.

References

Cameron A. G. W. (1963) (Editor) *Interstellar Communication: The Search for Extraterrestrial Life.* W. A. Benjamin, New York.

Claus R. and Wiebe E. (1974) Low-Noise Receivers: Microwave Maser Development. *JPL Tech. Rep.* 32–1526, Vol. XIX.

Cocconi G. and Morrison P. (1959) Searching for Interstellar Communications. *Nature* **184**, 844.

Dyson F. J. (1966) The Search for Extraterrestrial Intelligence. In *Perspectives in Modern Physics*, (Edited by R. E. Marshak), p. 641. John Wiley, New York.

Kardashev N. S. (1971) The Astrophysical Aspect of the Search for Signals from Extraterrestrial Civilizations. In *Extraterrestrial Civilizations, Problems of Interstellar Communication* (Edited by S. A. Kaplan), (Vnezemnye tsivilizatsii. Problemy mezhzvezdnoi svyazi); NASA TT F–631, of Izdatel'stvo "Nauka" Glavnaya Redaktsiya Fiziko-Matematicheskoi Literatury (Moskva 1969), Israel Program for Scientific Translations, Jerusalem.

Kardashev N. S. (1979) Strategy for the Search for Extraterrestrial Intelligence (NASA Translation NASA AS 77–98, of 0 strategii poiska vnezemnykh tsivilizatsiy), *Acta Astronautica* **6**, 33–46.

Kraus J. D. (1966) *Radio Astronomy*, McGraw-Hill, New York, 26 ff.

Oliver B. M. (1965) Thermal and Quantum Noise, *Proc. I.E.E.E.* **53** (5), 436–454.

Oliver B. M. (Oct. 1976) *The Rationale for the Water Hole*, XXVII Congress International Astronautical Federation, IAF–A–76–32.

Oliver B. M. (1978), in SETI (1978) below.

Oliver B. M. and Billingham J. (1972) *Project Cyclops: A Design Study of a System for Detecting*

Extraterrestrial Life. Prepared under Stanford/NASA/Ames Research Center 1971 Summer Faculty Fellowship Program in Engineering System Design, NASA CR 114445.

SETI (1978) The Search for Extraterrestrial Intelligence, is the final report of a two-year Science Workshop sponsored by NASA-Ames Research Center. Philip Morrison was chairman of the workshop made up of sixteen scientists from a variety of disciplines (1974–76). *NASA Publication* SP–419. See pages 11–31 of this volume for a portion of this study.

Appendix I

Extract from Complementary Document 9, by Mark A. Stull, in SETI (1978).

PROTECTION OF A PREFERRED RADIO FREQUENCY BAND

The Science Workshop, especially at its second, third, and fourth meetings, debated all considerations concerning whether the probability of finding signals from an extraterrestrial civilization is maximized in any particular frequency band. To the extent this problem requires knowledge of the motivation of such civilizations, it cannot be solved, but there are physical and philosophical arguments which imply that the frequency band between 1400 and 1727 MHz should have high priority for a search effort. Because of the sensitivity of any interstellar search system (ISS), it is very important that the only telecommunications services which operate in this band be those that will not cause harmful interference to an ISS. In 1979 a general World Administrative Radio Conference (WARC) will be held; it will allocate world-wide use of the radio spectrum and allocations made then are likely to determine spectrum usage for the remainder of this century. The Science Workshop recognized the importance of obtaining protection of the 1400 to 1727 MHz band for SETI use at the 1979 WARC and, to emphasize the need, adopted the following resolution.

STATEMENT ON THE REQUIREMENTS FOR PROTECTION OF AN INTERSTELLAR SEARCH SYSTEM (ISS) FROM RADIO-FREQUENCY INTERFERENCE

In recognition of the rapidly advancing national preparation for the 1979 general World Administrative Radio Conference (WARC), the Science Workshop adopts the following final statement of policy:

(1) There are important frequency bands for a search for radio signals from extraterrestrial intelligent civilizations. These are:

(a) 1.400 to 1.427 GHz

(b) 1.427 to 1.727 GHz

The 1.400 to 1.427 GHz band is of interest because interstellar transmissions may take place around the hydrogen line, while the 1.427 to 1.727 GHz band is located between the hydrogen and hydroxyl lines and lies near the minimum of the noise background. 1.400–1.427 GHz is currently allocated exclusively to the radio astronomy service and may be shared with it, while 1.427–1.727 GHz may be shared with services whose use will not cause harmful interference to the operation of an ISS.

(2) Existing radio telescopes are already being used to search for radio signals from extraterrestrial civilizations, while the feasibility of constructing a very large ground-based ISS has been established. The performance of any ground-based instrument will, however, be seriously degraded by radio-frequency interference, primarily from line-of-sight transmitters. The only identified alternatives to an Earth-based ISS are:

- A space-based ISS, and
- An ISS sited on the far side of the Moon.

Both of these are possible in the future, but we do not know at what cost. Furthermore, a space-based ISS, unless shielded at additional expense, remains vulnerable to interference from satellite and ground-based transmitters; while an ISS on the far side of the Moon is vulnerable to all transmissions originating beyond the lunar orbit. Thus, there exists a need for RFI protection. We strongly recommend:

(a) That the U.S. undertake immediate studies to determine detailed frequency protection requirements for an ISS, and submit the results of such studies to the 1977 Final Meeting of the International Radio Consultative Committee (CCIR) for inclusion in the supporting documents of the 1979 WARC, and,

(b) That the U.S. prepare and present to other administrations at the 1979 WARC a proposal which will include;

(i) Allocations of new satellite systems at frequencies outside the protected bands.

(ii) Appropriate frequency sharing criteria for uses compatible with the operation of an ISS.

(iii) Technical criteria for allowable spurious radiation from out-of-band uses.

(iv) Phase-out of interfering uses now operating in the protected bands.

Subsequent to the Science Workshop deliberations and during the preparation of the final report, the SETI radio frequency protection need received its first international recognition. Reproduced below is Addendum No. 1 to Volume II, XIIIth Plenary Assembly of the C.C.I.R., Geneva, 1974.

Note by the Director, C.C.I.R.

Subsequent to the publication of Volume II (Space Research and Radio Astronomy) of the documents of the XIIIth Plenary Assembly of the C.C.I.R., a new text relating to search for extraterrestrial life has been submitted for adoption by correspondence, in conformity with the provisions of No. 308 of the International Telecommunication Convention, Torremolinos, 1973.

It has received more than the twenty approvals necessary for its adoption by the Members of the I.T.U. and has therefore become an official Question of the C.C.I.R.

The text is as follows:

Question 17/2

RADIOCOMMUNICATION REQUIREMENTS FOR SYSTEMS TO SEARCH FOR EXTRATERRESTRIAL LIFE

The C.C.I.R., (1976)

Considering

(a) That many scientists believe intelligent life to be common in our galaxy;

(b) That electromagnetic waves are presently the only practical means of detecting the existence of intelligent extraterrestrial life;

(c) That it is believed to be technically possible to receive radio signals from extraterrestrial civilizations;

(d) That, although it is not possible to know the characteristics nor to predict the time or duration of these signals in advance, it is reasonable to believe that artificial signals will be recognizable;

(e) That, while an artificial radio signal of extraterrestrial origin may be transmitted at any frequency, it is technologically impractical to search the entire radio spectrum but the band searched should be sufficiently wide to make detection of a signal reasonably probable;

(f) that technological and natural factors which are dependent on frequency determine our ability to receive weak radio signals;

(g) That while the search for radio signals from extraterrestrial civilizations has already begun, more sensitive systems will be in use by the 1980s which could receive harmful interference from very weak man-made signals;

(h) That it is necessary to share the bands in which the search is conducted with other Services;

(j) That available technology will allow a search for these signals from the earth, from earth-orbit, and, eventually, from the moon and to minimize interference, certain locations on earth and in space may be preferred;

DECIDES that the following question should be studied:

(1) What are the most probable characteristics of radio signals which might be broadcast by extraterrestrial civilizations and the technical characteristics and requirements of a system to search for them;

(2) What are the preferred frequency bands to be searched and the criteria from which they are determined;

(3) What protection is necessary for receiving systems conducting a search for artificial radio signals of extraterrestrial origin;

(4) What criteria will make operation of a search system feasible in shared, adjacent and harmonically related bands of other Services;

(5) What is the optimum search method;

(6) What are the preferred locations, on earth and in space, for a search system?

Acta Astronautica. Vol. 6, pp. 129–143. Pergamon Press 1979. Printed in Great Britain

An observational program to search for radio signals from extraterrestrial intelligence through the use of existing facilities†

R. E. EDELSON‡
Jet Propulsion Laboratory, Pasadena, CA 91103, U.S.A.

Abstract—Approaches to the detection of radio signals from extraterrestrial intelligence (ETI) have, of necessity, circumscribed the vast parameter space which might be searched. Uncertainties in location, frequency, modulation, and reasonable power flux density have drawn arguments for huge antennas and philosophical approaches to frequency selection. Yet the observational limits within which ETI signals might await detection are largely undefined. The state of the art has advanced so that observations with existing antennas, equipped with modern data processors, can define limits over very broad frequency ranges in reasonable observational times. A modest program has been initiated to search for ETI signals from 1 to 25 GHz over 80% of the sky to flux levels from 10^{-19} to 10^{-23} W/m^2.

1. Introduction

From the ground-breaking work of Cocconi and Morrison (1959), through Projects Ozma (Drake, 1963) and Cyclops (Oliver and Billingham, 1973), and to the present day, there has been general recognition that, to operate within the limitations of extant technology, severe restrictions must be placed on the vast range of signal parameter space in any search for extraterrestrial intelligence (SETI). Even sanguine assumptions seeking detection from relatively nearby stars in the low-noise microwave window leave 20 GHz of the electromagnetic spectrum that must be examined to flux levels as low as 10^{-30} W/m^2 to truly encompass the range of signals that might be expected. With currently implemented devices, such undertakings would take centuries.

As a result, search activities have been of two types. The observational programs have concentrated on severely limited frequency windows, sometimes down to total bandwidth examinations of only 300 kHz or so. These windows have been chosen, not on the basis of physical argument alone, but through speculation as to the approach taken by the transmitting society. Studies, such as Project Cyclops (Oliver and Billingham, 1973), have discussed wider but still philosophically constrained frequency windows, while advocating beginning the implementation of antenna systems which might reach phenomenal size.

To be comprehensive, a search must examine certain basic dimensions of signal space: source location, source modulation, transmission frequency, and

†This paper presents the results of one phase of research carried out at the Jet Propulsion Laboratory, California Institute of Technology, under Contract NAS7-100, sponsored by the National Aeronautics and Space Administration.

‡Manager, Telecommunications Systems Section.

Table 1. General ranges of search parameters

Parameter	Range
Possible source location	4π steradians
Source modulation	Cooperative modulation to permit detection, thus either strong CW carrier with low data rate modulation or repeated high-energy pulses
Transmission frequency	1 to 100 GHz
Received power flux density	10^{-17} W/m^2 to 10^{-30} W/m^2 or less

received power flux density. Table 1 displays the general ranges that these parameters may assume based on physical argument alone, that is, those ranges that require little if any second guessing of the transmitting society's approach or prowess.

To date, the observational limits within which ETI signals might exist have remained largely undefined. This has been due, not to any lack of ardor on the part of search adherents, but rather to the need for technology breakthroughs to accomplish what is clearly a Brobdingnagian undertaking. The advent of superb techniques for electronic miniaturization and for the inexpensive replication of highly complex data-processing devices has today provided the means to examine the parameters of Table 1 in large scale with modest expense. Thus, we are able, for the first time, to observationally determine definitive limits to the presence of microwave emanations from other intelligent beings; for a substantial portion of the capability for finding these signals lies not in the application of ever larger antenna collecting areas, but rather in the use of millions or billions of simultaneous frequency channel observations combined with rapid and powerful data-processing techniques.

This paper sketches the employment of these techniques on existing antenna facilities in a program of modest expense and duration. The effort seeks to discover certain classes of extraterrestrial signals of intelligent origin while defining boundaries to the search problem throughout the range of interest. This program will simultaneously investigate radio astronomical phenomena of interest and define the background of environmental radiation for the purpose of determining physical limitations on SETI strategies and, generally, on the potential for deep-space communications. In addition, wider application is seen in the development of methods to deal with the environment of man-made radio frequency interference (RFI) as it affects these disciplines.

2. The search parameter space

Convincing arguments have been made (Oliver and Billingham, 1973; Morrison *et al.*, 1978) for devoting the primary SETI efforts to the detection of artificially generated electromagnetic radiation. There appears no other communication method that provides the economy of energy and speed of information exchange that can be accomplished through this medium.

Even restricting the search to the electromagnetic regime requires the searcher to examine multiple dimensions to seek out signals from cooperative beings. The parameters that must be determined are: location of the transmitting source, frequency or frequency range within which the source is transmitting, modulation or method of imparting information to that signal, and the power flux density that may be observed from the reception site.

Source location

Many observers have assumed that the location of transmitting sources will be associated with stars, particularly with stars of classes F, G, and K. This assumption is most natural, as our only example of intelligent life is indeed associated with such a star, the Sun. The limitation of search targets to these star classes is based primarily on three arguments. First, that stars of luminosities greater than F class are of insufficient age to have provided a breeding ground for intelligent species, given our beliefs as to the time scales required for Darwin–Wallace evolution. Second, that stars of luminosity less than spectral class K appear to be subject to violent coronal events which would tend to provide life with extremely hostile environments. Further, a planet close enough to such a star to lie in the so-called "ecosphere" would be in tidal lock with its primary, never presenting more than one face to the star, with disastrous effect on its atmosphere. Third, stars that have departed the main sequence have been precluded from search strategies because of the obvious difficulties in explaining how life might survive the catastrophic events concurrent with such departure.

Such arguments, while anthropocentrically appealing, can be disputed on the basis of naive speculation about the capabilities of societies 10^5 or more years our senior. Indeed, the adherents of searching only the F, G, and K spectral classes recognize that their arguments merely express likelihood, not certainty, and they remain satisfied to limit their search approaches because of the plethora of F, G, and K targets in relatively close proximity to the Sun. In a targeted search, the thorough examination of only these target stars, 2 million of which lie within 1000 light years of the Sun, is a task of sizable scope.

However, the fact remains that other stars may provide homesites to our galactic brethren. Indeed, the origin of signals cannot be limited merely to stellar systems. B. C. Murray (private communication), for example, has speculated that an advanced society might take advantage of the vast power output of interstellar masers to supply their information-bearing carrier, modulating such sources with relatively small input energy. Such sources of interstellar signals might not be obviously coupled to a stellar system and hence would remain unobserved in a targeted search.

In a similar vein, J. H. Wolfe (private communication) suggested that the most advanced societies may circle white dwarf stars. Surely any society that has survived a red giant expansion and the subsequent white dwarf collapse would be potent indeed. Such a scenario may prove heartening to those who contemplate with sorrow humanity's coming demise by the same mechanism about 5 billion years from now.

The realm of speculation is without limit and can provide diverting conversations around many a SETI fireside. However, such constructions should not be easily dismissed because scenarios can be thus arrived at which are not at variance with physical law as we understand it. Remember that the most conservative approaches to target selection rely on the extrapolation of an experiment with only one data point, ourselves.

What then is the alternative to limiting the search by particular targeting assumptions? As described below, a significant search can be undertaken which performs a whole-sky survey for likely radio sources. While such a survey can never approach the sensitivity of a targeted search within a reasonable search period, it can bound the power to received flux levels which require only modest assumptions as to transmitting society prowess.

Source modulation

Electromagnetic signals can be varied in many ways to provide the capability of bearing information. Human efforts in deep-space communications have applied the fundamental precepts of information theory to approach the theoretical limits of channel capacity for a given transmitting power level. Transmitting societies conducting interstellar communications would no doubt do the same, and most probably will have arrived at codes closer to the fundamental limits of Shannon's law than our own.

Such communication methods are of little use in initiating contact with a society. They rely on gaining communications advantages by the expansion of the signal over a wide bandwidth. As a result, no small portion of the spectrum would appear to have a meaningful signal within it. Indeed, the optimum transmission would be accomplished with signals which, without *a priori* knowledge of the code structure, would appear to be simply a minuscule addition to the natural noise background.

Thus, to attempt to search for signals of extraterrestrial intelligent origin, we must have the audacity to assume that the transmitting species are eagerly attempting to gain our attention. It is true that with very large collecting areas we might be able to eavesdrop on a society's internal communications (our own television communications, for example, still have very strong carrier components). However, the number of stellar systems within range of such an activity, even with phenomenally large receiving antennas, is quite small. A practical range for such detection is certainly not much more than 100 light years.

Being drawn to the inevitable conclusion that we must assume complicity on the part of the transmitting society in order to achieve an interstellar link, we can expect two types of modulation that a search may seek. First, a modulation scheme that provides a strong carrier component of very narrow width, continuously transmitted or, second, a transmission in which the signal is strong, and pulsed in a time scale that is short relative to a likely dwell time in a particular pointing direction.

We consider the former more likely since it represents the limiting case of the latter. Additionally, although powerful pulsed signals are known to exist in nature, very narrow band continuous signals are not. One of the important aspects of establishing the bounds on the SETI parameter space is to examine whether such signals are indeed precluded for natural phenomena, as we believe.

In any case, both types of signal may be detected with the same basic equipment complement by varying the data-processing techniques. The general approach is described below.

Transmission frequency

By transmission frequency, we refer to that frequency at which the continuous carrier may be received, or that frequency central to the pulse bandwidth received from the transmission of a powerful pulsed signal.

As noted above, for physical reasons we can circumscribe the search to the examination of the sky for artificial sources of electromagnetic radiation. Defining a search in this way hardly limits its magnitude at all. A requirement to search for signals over the entire electromagnetic spectrum represents such an incredible effort as to thoroughly frustrate any attempt at search activity. Thus we are again brought to examine the physical circumstances involved in the transmission of electromagnetic radiation to see if the reckoning of physics can further proscribe the search arena.

Generally, the methods of linear amplification used in the detection of signals in the radio regime of the spectrum imply a preferred spectral region that encompasses wavelengths of approximately 1 cm to 1 m—this has been called the microwave window. Laser communication adherents have argued that this spectral area is not unique, and that the methods of direct photon detection in the frequencies of the laser regime give better efficiencies for interstellar communication. Recently, B. M. Oliver (unpublished memorandum) demonstrated that such hopes are illusory, and that equivalent effects exist which result in a higher energy requirement for transmission than that realized in the microwave region. Since physical arguments provide a particular frequency band that is the most efficient for electromagnetic communications, it appears that the optimum strategy is to examine this band fully. Figure 1 provides a pictorial representation of this region for a spaceborne receiver and Fig. 2 for an Earth-based receiver, showing the relevant noise sources that impede such communications, and demonstrating the noise minimum that lies behind the choice of this as the likely region for extraterrestrial intelligent signal detection.

Project Cyclops, in seeking to define an even narrower region to ease the search burden, noted that for a society using omnidirectional transmitters for achieving contact, a physical phenomenon exists which might provide a preference for a particular narrow band in the microwave window. Since an omnidirectional source cannot be adjusted to remove the Doppler drifts caused by the relative motion between the Sun and the source, the received noise power in the optimum detection bandwidth is essentially scaled by frequency raised to the 1/2 power. When this operation is performed on the curves in Figs. 1 and 2, the situations emerge that are depicted in Figs. 3 and 4, respectively. This scaling considerably sharpens the minimum already described, and brings out for particular interest the frequency band dubbed the "water hole." This band, lying

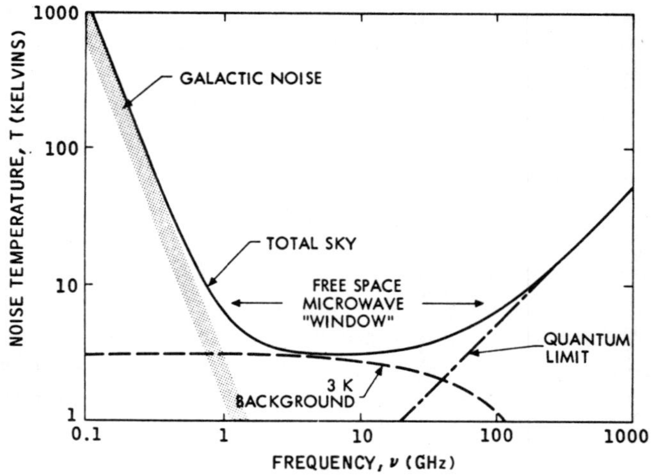

Fig. 1. Sky noise temperature for spaceborne receivers.

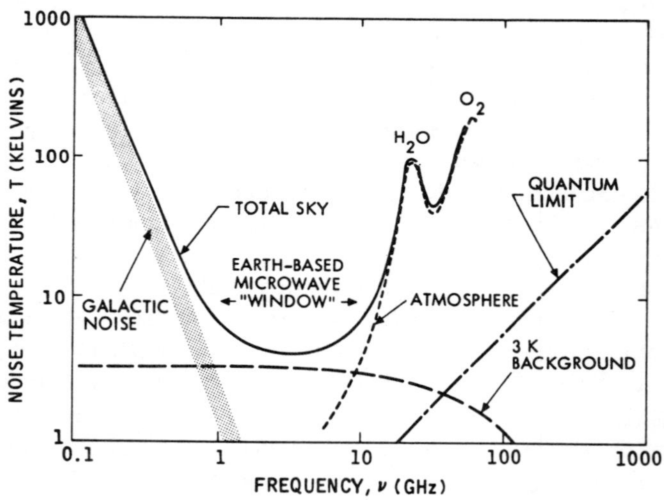

Fig. 2. Sky noise temperature for Earth-based receivers.

between the hydrogen and hydroxyl lines at approximately 1.4 and 1.7 GHz, remains of great interest for a SETI.

It is important to note, however, that the physics of the transmission process are less responsible for the selection of this band than the assumptions of (1) omnidirectional transmission and (2) the symbolism of the region for water-based life. Note, for example, that neither the H nor OH lines are actually produced by water, and a different philosophical choice might just as easily indicate the water line at 22 GHz as the preferred region for a search.

There is no doubt that the choice of search frequency is the parameter of most disparate opinion in the SETI community. While most agree that the

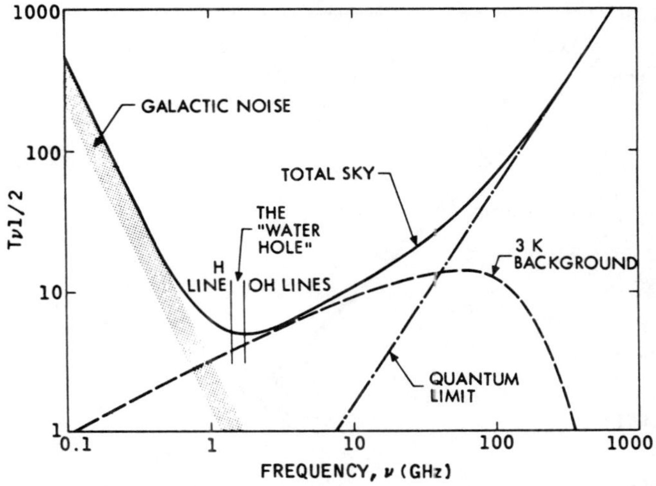

Fig. 3. Noise variable for omnidirectional source for spaceborne receivers.

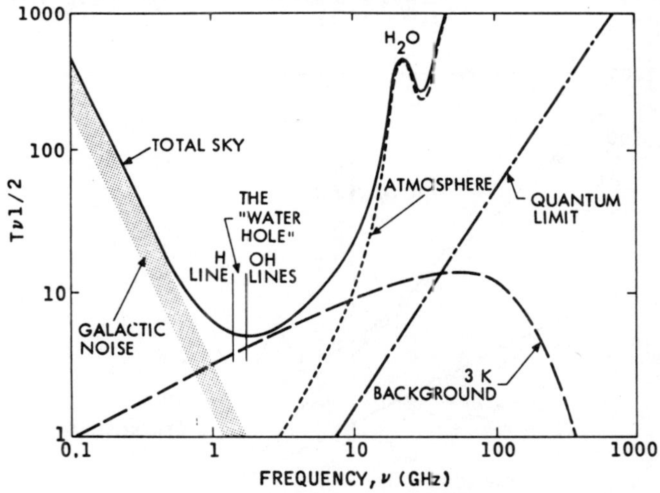

Fig. 4. Noise variable for omnidirectional source for Earth-based receivers.

"water hole" is an important band, many would prefer other regions as more likely search areas. Indeed, it may be argued that almost any frequency of radio astronomical interest is a likely region for seeking signals since the transmitting society is no doubt aware that radio astronomical instruments are the most likely receptors for their signals.

As a final speculation, let us note the situation depicted in Figs. 5 and 6. Here we have taken a view diametrically opposed to that of Project Cyclops, by assuming the existence of beamed transmissions. If a more advanced society than ourselves can winnow the possible sites for the growth of intelligent life more finely than we can, they may be capable of providing beamed trans-

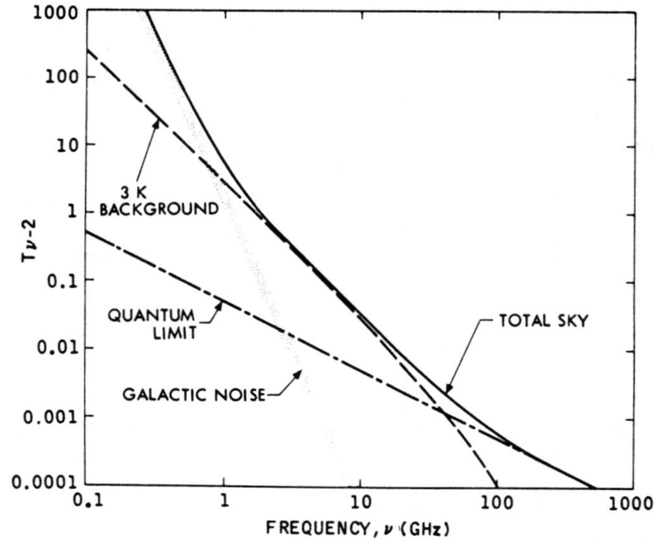

Fig. 5. Noise variable for beamed transmission for spaceborne receivers.

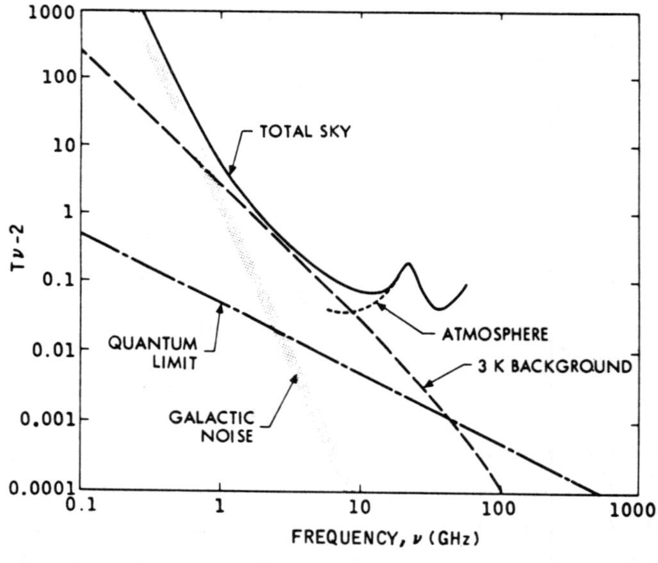

Fig. 6. Noise variable for beamed transmission for Earth-based receivers.

missions to those sites. Clearly, the power advantages are enormous for such a case. As has been pointed out many times (Sagan and Drake, 1975), the great telescope at Arecibo, Puerto Rico, equipped with only a few hundreds of kilowatts transmitting power, could communicate with an equal aperture halfway across the galaxy. The effective isotropically radiated power in such a case exceeds 10^{12} W. In the case of beamed transmissions, the signal advantage (for commensurate antenna surface accuracy and constant aperture) is proportional to frequency squared. Thus we may search for an appropriate frequency regime by examining the noise temperature divided by frequency squared. The results are shown in Figs. 5 and 6. Note that, for the spaceborne case (Fig. 5), there is no preferred frequency but rather the choice of optimum transmitting frequency is governed by the accuracy with which one can build a surface and point the resulting pencil beam, capabilities we cannot presume to estimate for a substantially more advanced society than ourselves. (Note that the argument breaks down as we approach optical frequencies where other noise sources predominate and the assumption of constant antenna area becomes untenable (B. M. Oliver, unpublished memorandum).)

For an Earth-based case (Fig. 6), site-dependent frequency preferences emerge below the water vapor resonance at 22 GHz and in the trough between the H_2O and O_2 resonances. The same philosophy devoted to the justification of the "water hole" could equally apply at these minima.

The net result of this frequency examination is to indicate that transmissions are plausible throughout the microwave window. Indeed, we might expect that signals at the high end of the region would be quite strong since the rationale for this area involves either beamed transmissions or the use of serendipitous sources such as interstellar masers. Based on physical arguments, we can conclude only that the limits for interstellar communications in frequency space are bounded by the microwave window.

Received power flux density

The careful reader will have noted that in each section above there has been some unavoidable discussion of the separately considered variables. In a way, that is the crux of the SETI problem. Assumptions made about any of the search parameters inevitably affect the ranges to which the remaining parameters can be examined.

What constitutes a level of power flux density which we might reasonably expect to be impinging on the Earth from an artificial source? Clearly, the answer is intimately entwined with the location (in range) of that source, the frequency transmitted, and the modulation assumed. Omnidirectional narrowband transmitters of 1 GW (the nominal transmitter envisioned in the Project Cyclops report) located 1000 light years from Earth would require our capability for signal detection to achieve the level of 10^{-30} W/m^2. Such a flux level is tens of decibels below that which existing antenna facilities could detect given even the ultimate data-processing gain that signal broadening through the interstellar medium permits. On the other hand, beamed transmissions with large antennas and reasonable transmitter powers would provide flux densities of 10^{-26} W/m^2 or more from the same range, a quantity which can be detected with today's technology. Figure 7 demonstrates the range of joint transmitter-receiver capabilities that would lead to a successful signal exchange. Clearly, the range of acceptable power flux densities covers many decades.

The startling fact is that, except over very limited frequency ranges, signals of flux densities as high as even 10^{-19} W/m^2 would not have been detected in the

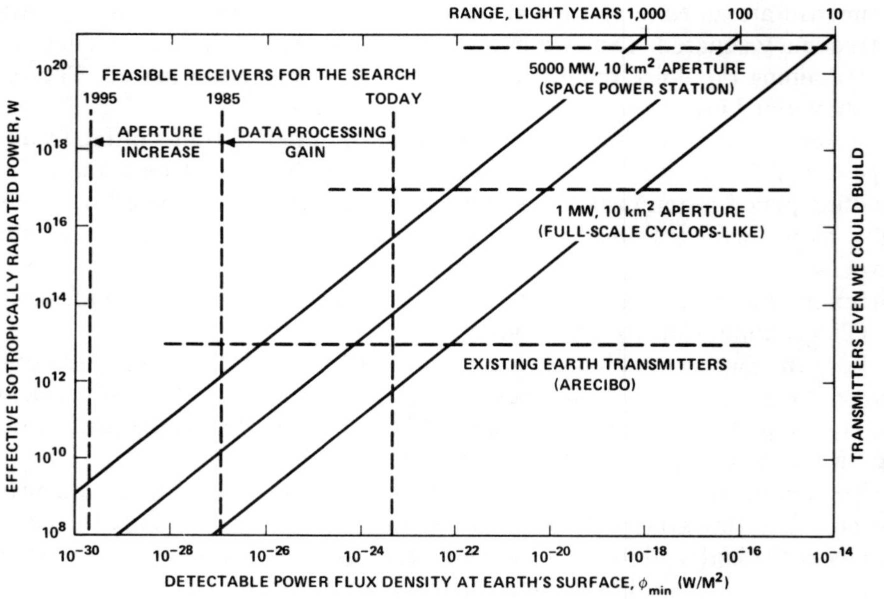

Fig. 7. Joint transmitter-receiver requirements.

daily observations now made in the microwave window. As noted in the SETI Science Workshops Final Report (Morrison *et al.*, 1978), an examination of the procedures followed in radio astronomical observations indicates that the coherent narrow-band signals that provide a prime candidate for interstellar signaling are actually discriminated against in the astronomers' data processing, so that detection of such signals could not be accomplished in the typical radio astronomical investigation. The reason for this discrimination is abundantly clear. Since the vast majority of signals of this type must arise from Earthly interfering signals, the data processing has been designed to preclude corruption of natural sources by such signals. Thus, in seeking to ameliorate the effects of our own radio emanations, we preclude the detection of signals from other intelligent beings. (Parenthetically, we note that for the same reason one of the fallouts of a broadband SETI activity such as we describe below must be a manner of identifying man-made interference and removing it from consideration in the data analysis associated with the search.)

We have established that the bounds on the parameter space within which a signal of extraterrestrial intelligent origin might exist remain largely undefined. What capability can be currently applied to determine these bounds, and how effective is it for this application?

3. Potential of existing facilities

A capsule evaluation of the capability of a radio astronomical facility to detect narrow-band signals may be achieved by determining the minimum power flux density that the system is capable of detecting.

Following Janssen and Gulkis (1977), we find that in order to accomplish detection, the ratio of the signal power, P_s, received from the source to the noise power standard deviation, σ, contributed by the observing equipment environment and the receiver electronics must exceed some threshold value C; that is,

$$\frac{P_s}{\sigma} \geq C \tag{1}$$

where C is chosen as a compromise that provides an acceptable probability of false detection while maximizing the probability of correct detection. A typical value of C for these circumstances is about 6.

For square law detection, with a channel of bandwidth B, observed for a time t, σ is found to be:

$$\sigma = \frac{P_N}{\sqrt{tB}} \tag{2}$$

where, for the microwave window, the noise power, P_N, is equal to kT_sB, Boltzmann's constant times the system noise temperature times the bandwidth of the channel. The minimum signal power that the system can detect is, then,

$$P_{\min} = CkT_s\sqrt{\frac{B}{t}}. \tag{3}$$

For an antenna of effective area A, the minimum power flux density is, simply,

$$\phi_{\min} = P_{\min}/A \tag{4}$$

Thus,

$$\phi_{\min} = \frac{CkT_s}{A}\sqrt{\frac{B}{t}}. \tag{5}$$

We note that $\phi_{\min}$ can be enhanced by any of the following: decreasing the channel bandwidth, increasing the observation time, or increasing the antenna area. Let us examine what might be accomplished by each of these actions.

System noise temperature

The theoretical limit to the system noise temperature in the microwave window is defined by the cosmic background temperature of approximately 3 K. Actually, system noise temperature is limited by current technology in maser amplification and in suppressed side-lobe antenna patterns. Temperatures as low as 13 K have been achieved in existing facilities, representing improvements by a factor of about 30 over the temperatures of systems used in early searches. However, the practical limit has been nearly achieved, and no more than an additional improvement by a factor of about 2 can be expected through further development of amplifiers, low side-lobe antenna designs, and possibly space-borne antennas.

Observation time

Sensitivity increases slowly with an increase in the observation time. As a practical matter, the desirable trend is to decrease observation time, as any search strategy is predicted on a large number of individual observations. Thus, the avenue of increasing observation time to increase sensitivity is foreclosed to any substantial search.

Antenna area

An increase in antenna area is the one avenue to increasing the sensitivity of a search system without limit, particularly for spaceborne systems where atmospheric corruption of directivity is not a factor. However, the cost of increased aperture is very large, thus precluding the use of this avenue until the capabilities of the existing facilities are exhausted.

Channel bandwidth

Channel bandwidth can be decreased to obtain increased system sensitivity. The physical limit to this method is found not in the hardware but rather in the dispersive medium through which the signals pass. Although channel bandwidths as low as 10^{-5} Hz are possible with hydrogen maser clocks, the interstellar medium will probably degrade any transmitted signal to a bandwidth of perhaps 10^{-3} Hz. Even this bandwidth is probably small compared to uncertainties in providing compensation for stellar relative motion. Finally, data modulation of the signal will further widen the received spectrum. Thus, the anticipated signal bandwidth lies in the range 0.1–10 Hz. Decreasing the receiver bandwidth below the signal bandwidth results in a rapid loss of received power, thus decreasing the capability for detection.

Extant radio astronomical facilities have the capability of achieving bandwidths in this region, although the minimum examined for astronomical purposes is typically on the order of 1–10 kHz. Thus, by narrowing the bandwidth, we might anticipate an improvement of as much as 300 in the sensitivity of the existing systems.

This potential sensitivity increase has not been achieved for two reasons. First, there are no known astrophysical phenomena observable at bandwidths of much less than 1 kHz. Second and most important, the existing equipment consists of multichannel spectral analyzers of no more than a thousand or several thousand spectral bins.

Thus, significant improvement in sensitivity by decreasing bandwidth through the use of these devices would result in such a narrow total frequency examination as to preclude any realistic detection attempt.

Developments are now underway to provide multichannel spectral analyzers based on state-of-the-art electronic technology. These devices, which will permit the achievement of perhaps a factor of 100 improvement with existing antennas, will begin to be available, under current plans, in about 2 years. The early devices, designed particularly for the search for extraterrestrial intelligence, will achieve 10^6 channel capability over a bandwidth of up to 300 MHz for a cost of about 90¢/channel. Future developments are already anticipated which will permit the construction of devices with 10^9 channels or more at costs in the range of 1¢/channel.

Simultaneously, developments in low-noise amplification, having nearly reached the physical limits to noise temperature reduction, are now concentrating on the achievement of much greater bandwidth and tunability. Concurrent with the development of the multichannel spectral analyzers, and required for their effective application, development is proceeding on a low-noise amplifier (noise temperature of 4 K or less for the device itself) which will have an instantaneous bandwidth of 300 MHz and a tuning range of from 1 to 25 GHz.

These analyzers and their companion low-noise receivers are the key to a comprehensive search at modest cost, as they will permit the use of existing antenna facilities, improved in sensitivity by one to two orders of magnitude, for costs which are quite modest when compared to the only other alternative, that of increasing the antenna aperture area by the same factor. Through the use of these devices, a viable search can be launched.

4. A modest proposal

The National Aeronautics and Space Administration is now in the formative stages of a search for evidence of extraterrestrial intelligence. As part of this search, both the Ames Research Center and the Jet Propulsion Laboratory are embarking on observational programs, cooperatively developing the technology required for their simultaneous accomplishment. The Ames program is primarily an examination of the signal space in the dimension of power flux density, with limited spatial and frequency coverage. The JPL approach will explore the dimensions of location and frequency. The latter program is described below.

Program description

The primary emphasis of this observational program is to carry out a SETI over a broad frequency range and with large sky coverage by utilizing existing antennas and state-of-the-art microwave receivers and spectrum analyzers. The observations will encompass 80% of the sky over the entire frequency range of from 1 to 25 GHz with a sensitivity limit which varies from 10^{-21} W/m^2 at the lowest frequencies to 10^{-19} W/m^2 at the higher frequencies. At a few selected frequencies of SETI and scientific interest, including the "water hole," this sensitivity limit will be lowered to 10^{-23} W/m^2 over 80% of the sky. In addition, a 500-star targeted search will be performed in the "water hole," achieving flux levels as low as 10^{-24} W/m^2.

A key element of the program is the constraint to accomplish significant results within a time frame of 2 years of development and 5 years of observational activity. Analysis has shown that in the performance of a search, data of significant interest to radio astronomy will be gathered (Morrison *et al.*, 1978). The program will yield data required for the design and operation of future, more sensitive SETI programs, should they be required. A characterization and understanding of radio frequency interference (RFI) over a broad frequency range will be obtained which will be valuable for SETI and for the evaluation of the limits to deep-space communications generally. Associated problems of signal discrimination, pattern recognition, data handling, and fully automatic systems operations will be addressed.

Program goals

The program goals may be broadly stated as follows:

(1) Detect the presence of radio signals of extraterrestrial intelligent origin.
(2) Without compromising the search, tailor the program to gather radio astronomical data of scientific interest.
(3) Build an observational and technological framework on which future, more sensitive SETI programs could be based.
(4) Develop approaches for identifying, cataloging, and characterizing interference signals (RFI) to permit identification of sought-for data.

Anticipated achievements

The constraint that significant results be possible within the specified 5-year observation places certain restrictions on the possible strategies. Significant means that substantial contribution be made to proscribing the range of the unknowns for SETI examination. Janssen and Gulkis (1977) have shown that constraining a search activity in this way has certain implications on the extent to which fruitful examination can proceed.

(1) When broadband frequency and whole-sky surveys are contemplated, the limiting sensitivity is determined by data-processing capability. To achieve an all-sky survey with a single antenna over the frequency range 1–25 GHz in 5 years or less requires the use of an average beamwidth of at least 3°.

(2) Large antenna apertures (narrow beamwidths) are desirable when data processing ceases to be a limitation. This implies either a survey that is limited in frequency or spatial coverage or the application of more facilities.

Thus, a strategy has evolved which utilizes a series of antennas, differing in beamwidth and specifically chosen to accomplish the program goals within the 5-year time frame. A wide-band flux background will be established by using a constant beamwidth antenna (15° or less) that scans the entire available sky at the sidereal rate. Particular frequency ranges of SETI and scientific interest will be examined to a greater sensitivity by using a 26-m antenna such as those of the NASA Deep Space Network (DSN) and, at the highest frequencies, by a specially equipped DSN 9-m dish. Figure 8 exhibits, as a function of frequency, the planned flux level limits which will be established by this program.

Program planning

The details of the frequency search, specific frequencies to be examined at higher sensitivity, and the time phasing of the search are areas that still require definition. The most vital area to be examined is the design of the data-processing algorithms. There is clearly a fine balance between worthwhile data and noise in such a system. The algorithm selection that distinguishes between these signals will require careful and detailed planning as well as observational experience. For this reason, JPL held a symposium in 1977 to assist us in the planning and experimentation required to accomplish the program goals. The

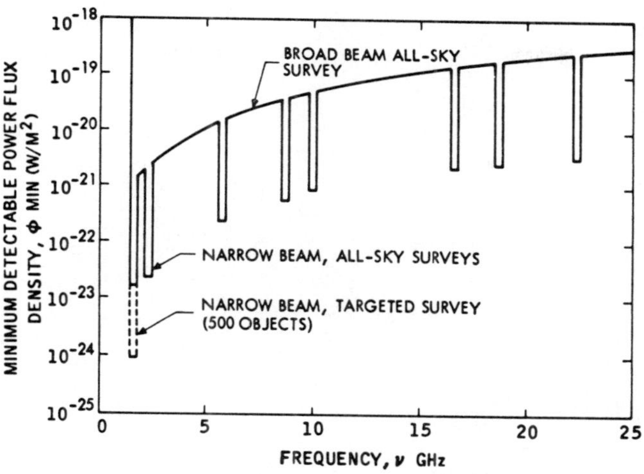

Fig. 8. Anticipated power flux density limits.

technology developed for this program will be offered to available facilities that wish to participate in this fascinating exploration.

5. Conclusion

The technology of data processing and microwave amplification has reached a point that allows, within 2 years, a SETI program of unprecedented sensitivity and breadth using existing antenna facilities. The Jet Propulsion Laboratory is now in the process of planning and designing such a search and welcomes the participation of the scientific community in this effort.

References

Cocconi G. and Morrison P. (1959) Searching for Interstellar Communications. *Nature* **184**, 844–846.

Drake F. (1963) Project Ozma, In *Interstellar Communication* (Edited by A. G. W. Cammeron), Banjamin Press, New York.

Janssen M. A. and Gulkis S. (1977) Parameterizing an All-Sky, All-Frequency Microwave Search for Narrow-Band Radiation of Extraterrestrial Origin. Paper presented at SETI Workshop, Jet Propulsion Laboratory, June 29–30 1977.

Morrison P., Billingham J. and Wolfe J. (1978) *SETI: The Search for Extraterrestrial Intelligence*, NASA SP-419.

Oliver B. M. and Billingham J. (1973) *Project Cyclops*: A Design Study of a System for Detecting Extraterrestrial Intelligent Life, NASA CR 114445, Revised edition.

Sagan C. and Drake F. (1975) The search for extraterrestrial intelligence, *Scientific American*, **232**(5), 80–89, May.

Acta Astronautica. Vol. 6, pp. 145–162. Pergamon Press 1979. Printed in Great Britain

An experiment protocol for a search for radio signals of extraterrestrial intelligent origin in the presence of man-made radio frequency sources†

R. E. EDELSON‡
Jet Propulsion Laboratory, Pasadena, CA 91103, U.S.A.

Abstract—The major difficulty encountered in the conduct of a comprehensive search for radio signals of extraterrestrial intelligent origin is the vast range of fundamental parameters that must be examined to make significant inroads on the plausible signal regime. This problem is aggravated by the almost certain occurrence of mistaken detections caused by stochastic processes and by man-made radio frequency emissions. Each such mistake, unless quickly and unambiguously characterized as a false alarm, will require repetitive observations during the conduct of the program. This wasted time will, in effect, reduce the sensitivity achievable in a given search duration This paper examines the implications of false alarms on the search system design, performance, and application and, in particular, on the choice of the data-processing approach. We show that, by an approach that limits the requirement for repeated observations, the degradation caused by false detections can be constrained to less than approximately 40%, whereas, without this restriction, the degradation might be severe. By modeling the mistaken detections caused by man-made devices as a Poisson process, we can place an upper bound on the probability of false alarms as a function of the detection threshold and of the number of interferors in the hemisphere. The understanding developed by these considerations suggests an experiment protocol consisting of a set of sequentially applied criteria to limit the range of parameters that need to be considered in declaring a false alarm, thus decreasing the time required to make this declaration and enhancing the performance of the search.

I. Introduction

THE PRIMARY emphasis in defining a strategy to search for radio signals of extraterrestrial intelligent origin must be to make significant inroads upon the vast regime opened by plausible hypotheses about the sender's approach. Assuming that we will need to explore a substantial fraction of this regime in testing these hypotheses, it quickly becomes obvious that the extent of the unknowns we face in frequency, location, received flux, and modulation precludes any presumption of an early detection. Modest programs will require years to accomplish their objectives, and large system proposals have discussed decades or even centuries of search activity. Thus, even a small fraction of

†This paper presents the results of one phase of research carried out at the Jet Propulsion Laboratory, California Institute of Technology, under Contract NAS7-100, sponsored by the National Aeronautics and Space Administration.

‡Project Manager, Search for Extraterrestrial Intelligence; Manager, Telecommunications Systems Section.

search time allocated to the pursuit and resolution of false indications of detection will translate into months or years of fruitless effort, while unnecessarily stringent restrictions to reduce this fraction risk a missed detection. When searching for a needle in a haystack, it is foolish to wear gloves.

This paper addresses some of the concerns faced by the designer of one kind of search system, a system to detect narrow-band radio signals of extraterrestrial origin and distinguish them quickly and unambiguously from those that we generate. By applying a rather simple experiment protocol, a strategy can be developed for handling these false alarms. Generally, search efforts must achieve their objectives within a constrained search duration, and false alarms must degrade the sensitivity to which the search might otherwise be conducted. A rapid means of false-alarm detection, preferably carried out within the uninterrupted conduct of the experiment, can minimize this degradation. The method described here is one way to approach this task.

II. General relations for a microwave search

In this section, we summarize and slightly expand the work of Janssen and Gulkis (1977), who derived parametric relations applicable to a microwave search of constrained duration which employs fast Fourier transform spectrum analyzer receivers, with sensitivity enhancement by spectrum accumulation and detection by a threshold criterion.

Spectrum accumulation is equivalent to the condition that $tB \gg 1$, where t is the accumulation or integration time and B is the single-channel bandwidth of the spectrum analyzer. Under these conditions, with a Gaussian white noise input, the statistics of a single-channel output are approximately Gaussian with mean $\bar{p}$ and standard deviation σ:

$$\bar{p} = kT_sB \tag{1}$$

and

$$\sigma = \bar{p}/\sqrt{(tB)} \tag{2}$$

where k is Boltzmann's constant and T_s is the system noise temperature.

In order to declare a detection, we require that the output, p_s, of a channel exceed σ by a factor α, where, as shown below, α is determined by the allowable probability of false alarm. Thus, detection is announced when

$$p_s \geq \alpha\sigma$$

or, substituting from (1) and (2), the minimum detectable signal power p_{sm} is

$$p_{sm} = \alpha kT_s\sqrt{(B/t)}$$

and the minimum detectable flux, ϕ_m (in W/m^2), is just

$$\phi_m = (\alpha kT_s/A_e)\sqrt{(B/t)} \tag{3}$$

where A_e is the antenna effective area, generally given by ηA (antenna efficiency times physical area).

Now let us assume that we observe a spectral region NB (N is the number of channels of bandwidth B that the spectrum analyzer processes) centered at a frequency f_i where $NB \ll f_i$ (this condition simply says that the system characteristics are essentially constant over the observed band). Then, if we wish to observe M_i directions or a fraction $g = M_i\Omega_i/4\pi$ of the sky, where Ω_i is the usable antenna beam solid angle† at frequency f_i, and if each direction is of equal *a priori* interest (equal integration times), then this requires a time T_i where

$$T_i = M_i t$$

or, from (3),

$$T_i = M_i B(\alpha k T_s / A_e \phi_{mi})^2.$$

Then, for an observation of L bands NB wide, the search duration can be found from

$$T = \sum_{i=1}^{L} T_i = B(\alpha k T_s)^2 \sum_{i=1}^{L} \frac{M_i}{(A_e \phi_{mi})^2} \tag{4}$$

where, for simplicity, T_s is assumed constant with frequency over the range $LNB = f_L - f_1$.

Let us look at several special cases.

Case I. Targeted search, constant aperture and flux, many contiguous bands

For this case, we have the following conditions:

1. $M_i = M = \text{const}$
2. $A_e \approx \text{const}$ for an antenna capable of viewing the assumed frequencies
3. $\phi_{mi} = \phi_m = \text{const}$
4. $(f_L - f_1)/NB = L \gg 1$.

Then,

$$T = M(\alpha k T_s / \phi_m A_e)^2[(f_L - f_1)/N]$$

and the flux that can be obtained is simply

$$\phi_m = (\alpha k T_s / A_e)\sqrt{(M(f_L - f_1)/NT)} \tag{5}$$

Case II. Sky survey, constant aperture and flux, many contiguous bands

For this case, we have the following conditions:

1. $M_i\Omega_i/4\pi = g = \text{const}$
2. $A_e \approx \text{const}$

---

†The usable antenna beamwidth refers to that portion of the main lobe selected as defining the spatial extent of an observation. Depending on the sensitivity sought and the nature of the search strategy (*a priori* target or search for sources with unknown locations), this might be the 1-, 3-, or 10-*dB* beamwidth.

3. $\phi_{mi} = \phi_m = \text{const}$
4. $(f_L - f_1)/NB = L \gg 1.$

Then,

$$T = 4\pi g B(\alpha k T_s/A_e\phi_m)^2 \sum_{i=1}^{L} \frac{1}{\Omega_i}.$$

Now

$$\Omega_i \approx \epsilon \frac{c^2}{A_e f_i^2}$$

where ϵ is a constant that depends on the specific antenna applied and c is the speed of light, so that,

$$T = (4\pi g B/\epsilon A_e)(\alpha k T_s/c\phi_m)^2 \sum_{i=1}^{L} f_i^2$$

and, for L large,

$$T \approx (4\pi g/3\epsilon A_e N)(\alpha k T_s/c\phi_m)^2(f_L{}^3 - f_1{}^3)$$

and the obtainable flux is,

$$\phi_m = (\alpha k T_s/c)\sqrt{[(4\pi g/3\epsilon A_e N T)(f_L{}^3 - f_1{}^3)]} \qquad (6)$$

Case III. Sky survey, constant beamwidth, constant integration time, many contiguous bands

For this case, we have the following conditions:

1. $\Omega_i = \Omega = \text{const}$
2. $M_i = M = 4\pi g/\Omega = \text{const}$
3. $A_e = \epsilon c^2/\Omega f_i^2$
4. $\phi_{mi}/f_i^2 = (\alpha k T_s \Omega/\epsilon c^2)\sqrt{(B/t)} = \text{const} = \phi_{m1}/f_1{}^2$
5. $(f_L - f_1)/NB = L \gg 1$

Then,

$$T = 4\pi g \Omega B(\alpha k T_s/\epsilon c^2)^2 \sum_{i=1}^{L} (f_i^2/\phi_{mi})^2$$

or,

$$T = 4\pi g \Omega(\alpha k T_s f_1{}^2/\epsilon c^2 \phi_{m1})^2[(f_L - f_1)/N]$$

and now the obtainable flux is a function of frequency as,

$$\phi_{mi} = (\alpha k T_s f_i^2/\epsilon c^2)\sqrt{[(4\pi g \Omega/NT)(f_L - f_1)]} \qquad (7)$$

We note that in all these cases (5)–(7), an increase in M_i, or equivalently g, results in an increase (degradation) in ϕ_{mi} by a squareroot relationship.

III. Importance of false alarms

Generally, we anticipate false alarms, the indications of a detection when in fact none has been made, to arise from two sources. The first, and easiest to deal with, are those signals that exceed the detection threshold through random noise fluctuations. These are completely statistically determined by the nature of the noise input and by the detection process, and the expected value of the number of false alarms is readily calculated.

The second type is caused by the detection of a narrow-band signal ($B_{\text{signal}} < B$) which is, in fact, present but is not of extraterrestrial intelligent origin. For bandwidths $B < 1$ kHz or so, the only known source of such signals is man himself. In what follows we examine the probability of this occurrence.

There is also a third type of false alarm. Because of the presence of radiating molecular and atomic species in space and in the Earth's atmosphere, the detected spectrum is not flat in frequency. To avoid classifying the peaks and valleys of these signatures as detected signals, the spectrum is first "normalized" by the application of either our expectation of the spectral shape in these regions or an average of measured spectra. Any departure from this prestored signature is of *a priori* interest and therefore such "detections" will not be classified as false alarms since they will be examined for the ancillary radio astronomical aspects of the search activity.

In considering a spectrum analyzer detection device with a large number, N, of independent channels, the false-alarm probability is an important parameter in the determination of system performance. Selection of the channel threshold is a compromise between the probability of false alarm and the probability that a weak, but genuine, detection will be missed—the lower the former, the higher the latter. If we consider the probability of obtaining few false alarms for the spectrum analyzer as a whole, we are driven to select an exceedingly low probability of false alarm for the individual channel. To see this, let us assume that the individual channel probabilities of false alarm are independent and denote them by p_{fj}, where $j = 1, 2, \ldots, N$. Let us further assume that our normalization process is sufficiently good, and the man-made radio environment sufficiently uniform for small changes in frequency ($\Delta f/f \ll 1$), as to have $p_{fj} = p_f$, a constant for each spectral region NB that we examine. Then the probability of false alarm in a single observation which covers the entire frequency range NB for one beamwidth of antenna observation is

$$P_f = 1 - (1 - p_f)^N. \tag{8}$$

Figure 1 plots this quantity as a function of p_f for several values of N.

What determines an acceptable value for P_f? Let us define a false alarm for the system as the event that one or more channel false alarms are observed in the spectral region NB. Further, let us assume that the advent of each false alarm requires that the system be repointed for a second look at that portion of sky, *but not more than two looks*.

Now, the expected number of false alarms, N_{fi}, is just

$$N_{fi} = M_i P_{fi}$$

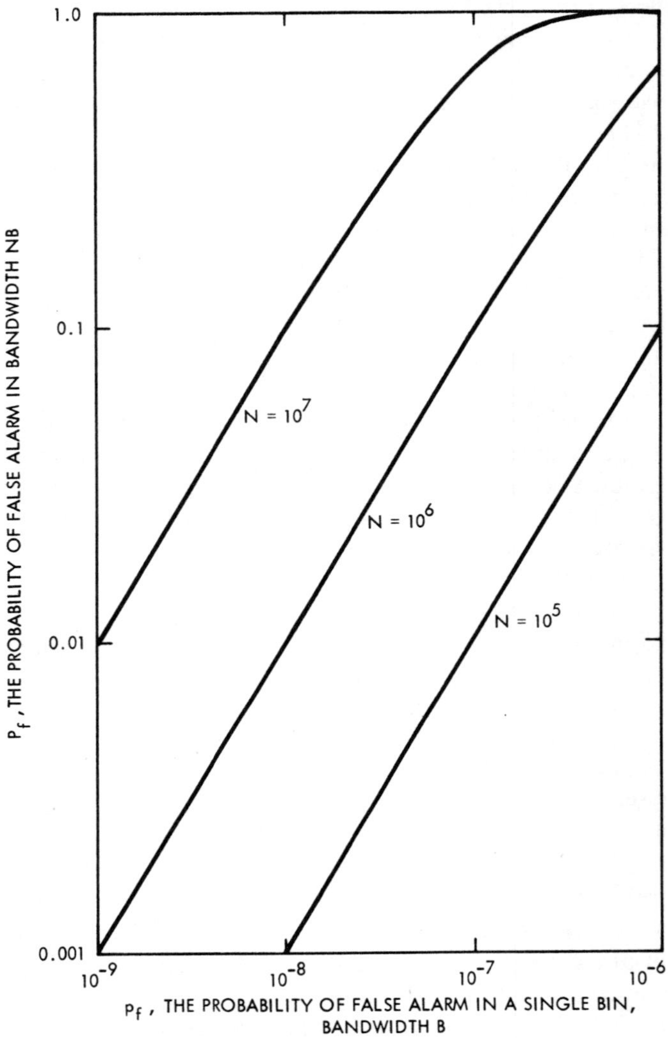

Fig. 1. Probability of false alarm for observation of a single sky direction over bandwidth *NB*.

where M_i is the number of target directions for our search of *NB* and P_{fi} is assumed independent of pointing direction (index i indicates that we are considering the problem at a center frequency f_i). Then the number of observations required for coverage of the frequency swath *NB* will be multiplied by a factor $(1+P_{fi})$. Recalling the analysis in Section II, eqns (5)–(7), and noting that each M_i is simply multiplied by $(1+P_{fi})$ to include the expected number of antenna pointings required, we obtain the modified flux equations:

Case I:

$$\bar{\phi}_m = \phi_m\left[\frac{1}{L}\sum_{i=1}^{L}(1+P_{fi})\right]^{1/2} \tag{5'}$$

Case II:

$$\bar{\phi}_m = \phi_m\left[\frac{3(f_L - f_1)}{L(f_L^3 - f_1^3)}\sum_{i=1}^{L}(1 + P_{fi})f_i^2\right]^{1/2} \tag{6'}$$

Case III:

$$\bar{\phi}_{mi} = \phi_{mi}\left[\frac{1}{L}\sum_{i=1}^{L}(1 + P_{fi})\right]^{1/2} \tag{7'}$$

where the superscript bar indicates the expected value of minimum achievable flux obtained in a fixed time duration T.

Note that in cases I and III, where the number of aiming directions in the sky is constant, the effect of false alarms on the expected value of achievable flux is independent of frequency. In case II, where the number of aiming directions increases by the frequency squared, the expected value of achievable flux is degraded in a manner which is weighted by the same factor.

In all cases, we may write $\bar{\phi}_{mi} = D\phi_{mi}$ and we can define a degradation due to false alarms as,

$$\text{Degradation } (dB) = 10 \log(\bar{\phi}_{mi}/\phi_{mi}) = 10 \log D.$$

As one more special case, consider the condition $P_{fi} = P_f$, independent of the observation frequency. Then, all cases reduce to,

$$\bar{\phi}_{mi} = \phi_{mi}(1 + P_f)^{1/2}$$

and,

$$\text{Degradation } (dB) = 10 \log(1 + P_f)^{1/2} = 5 \log(1 + P_f) \tag{9}$$

These equations lead to a degradation ranging from 0 dB, or no degradation ($P_{fi} = 0$), to 1.5 dB ($P_{fi} = 1$), that is, since we have foreclosed the possibility of more than two looks, we can do no more harm than a 40% degradation in the achievable flux density, even when interference is certain. The importance of ascertaining that a false alarm has occurred in no more than two looks is seen to be substantial. Without this ability, the degradation can be infinite, while with it the degradation is not only tolerable, but of the same order as other uncertainties in the system performance of untried hardware. The question now becomes one of ascertaining what is necessary to achieve this criterion. To do this, we first investigate the parameters that specify P_{fi} to show that, with an appropriate data-gathering procedure, the probability of requiring more than two looks is small. From this information we lay out an experiment protocol.

IV. Probability of false alarm

We begin the evaluation of P_f by computing the probability of false alarm in a single data channel, p_{fj}, for a single pointing direction of the antenna. In this derivation, we assume that there are no preferred directions for the occurrence of interference or noise fluctuations over the observed celestial sphere. As virtually all observations are taken at elevation angles substantially greater than zero, this assumption amounts to neglecting reception of Earth-based transmitters through atmospheric "bounce" phenomena and calibrating noise temperature variation with pointing direction.

We have chosen a simple threshold test to determine a detection. We shall define a detection, whether real or in error, by the condition:

$$p_{sj} \geq p_{smj}$$

and a nondetection by (10)

$$p_{sj} < p_{smj}$$

We define four possible events in the measurement of flux in a single channel by their probabilities as:

p_{oj} = probability of no detection, *whether or not there actually was a signal*
p_{Ij} = probability of a detection caused by a man-made interfering signal
p_{Nj} = probability of a detection caused by a noise fluctuation
p_{Ej} = probability of a detection caused by a signal of extraterrestrial intelligent origin

Then $p_{oj} + p_{Ij} + p_{Nj} + p_{Ej} = 1$. Here we have assumed that the four events are independent, which will be the case for a choice of high threshold ($\alpha \gg 1$) (i.e. the joint probability of a noise burst and an interfering signal occurring simultaneously in the same channel is very small).

The parameter p_{fj} then is described by

$$p_{fj} = p_{Ij} + p_{Nj}. \tag{11}$$

From (8), for N channels with a normalized spectrum,

$$P_f = 1 - (1 - p_I - p_N)^N \tag{12}$$

(p_{Ij} and p_{Nj} independent of the specific channel in the band NB). Note that we have assumed that $p_{Ej} \ll 1$. If this is not the case, our concern for false alarms is misplaced.

Now let us evaluate p_I and p_N.

Probability of a noise-related false alarm, p_N

The statistics of false alarm due to the passage through threshold of a Gaussian input signal processed by a square-law (power) detector such as a spectrum analyzer have been extensively examined (see Oliver and Billingham, 1973). Here we simply state the results for a Gaussian output approximation, valid for $tB \gg 1$. The term p_N is the probability that, due to system Gaussian noise, p_s exceeds $p_{sm} = \alpha k T_s \sqrt{(B/t)}$. As above, the power may be approximated as a Gaussian distributed random variable with mean kT_sB and standard deviation

$\sigma = kT_s\sqrt{(B/t)}$. Thus,

$$p_N \approx \frac{1}{\sqrt{(2\pi)}}\int_\alpha^\infty e^{-u^2/2}\,du \equiv (1/2)\mathrm{erfc}(\alpha/\sqrt{2}). \tag{13}$$

Note that the probability of requiring more than two looks to verify the occurrence of a noise-related false alarm is simply the probability of two consecutive strikes in the same channel or p_N^2.† Reasonable values of p_N are so small, due to the very large exponent of (12), that the need for more than two looks is negligible.

Probability of an interference-related false alarm, p_I

What constitutes an interfering source?

The flux received from an interferor can be written:

$$\phi = P_e/4\pi r^2$$

where P_e is the EIRP (equivalent isotropically radiated power) and r is the range from receiver to source. Generally, for SETI we seek a ϕ_m of 10^{-19} W/m^2 or less. What EIRP is required to obtain these levels from an offending source? Figure 2 plots the P_e that would generate a flux ϕ from a distance r. It is clear that even extremely small terrestrial emanations can pose a severe interference problem for a search. However, the definition of an interfering source in the context of the effect on SETI second looks can be confined significantly, particularly for search strategies that require a very "dense" search in direction.

We choose to divide man-made signals into two classes, those sufficiently strong to enter through the side lobes of the antenna and those that can only be detected near the gain peak of the main beam, the antenna boresight. The former may be easily discriminated by their continued presence in the band despite motion of the antenna to a new aiming point, and such a procedural step will be included in our experiment protocol. The latter class is responsible for the false-alarm possibility. Consequently, we confine this discussion to those signals strong enough to be detected only if they emanate from a source seen within a small angle around the boresight of the antenna.

Therefore, let us define p_{Ij} more closely as the probability that a flux ϕ_j of bandwidth $B_I \leq B$ derived from a man-made source impinges upon our detection system with

$$\phi_{mj} \leq \phi_j \leq \phi_I \tag{14}$$

†Note that we have neglected here considerations of source Doppler drift for repeated observations. If the observing bandwidth is very small ($\lesssim$ 1 Hz), it becomes necessary to consider channels adjacent to that of the first signal occurrence as being of a corroborating nature for a true detection. While such considerations are important for the actual search algorithm, they alter only the details of the implementation and not the basic thrust of the argument presented.

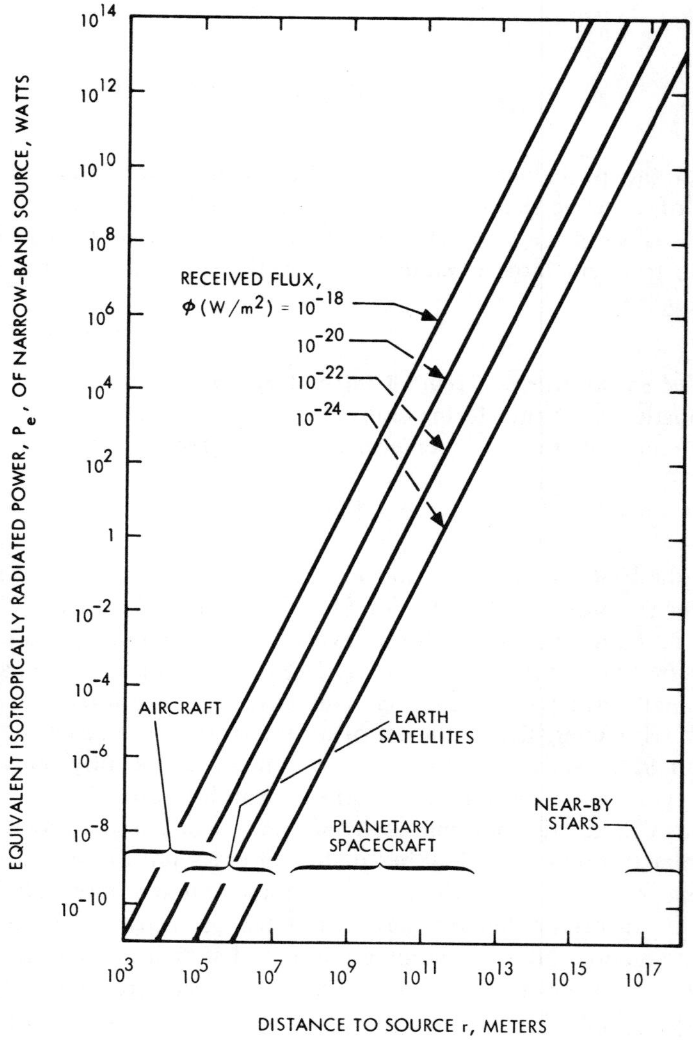

Fig. 2. EIRP required from distance r to create flux ϕ.

where ϕ_I is that flux above which the signal would be observed in the channel despite retargeting to the next search objective.

Let us examine the case of an interference source at an angle θ_I from the antenna boresight which has caused a false detection in one observation. If θ is the conical half-angle measured from the boresight (see Fig. 3) and assuming, for simplicity, that the antenna power pattern can be characterized by a monotonically decreasing gain function $G(\theta)$, then, recalling that by definition,

$$G(\theta) = \frac{4\pi f^2 A_e(\theta)}{c^2}$$

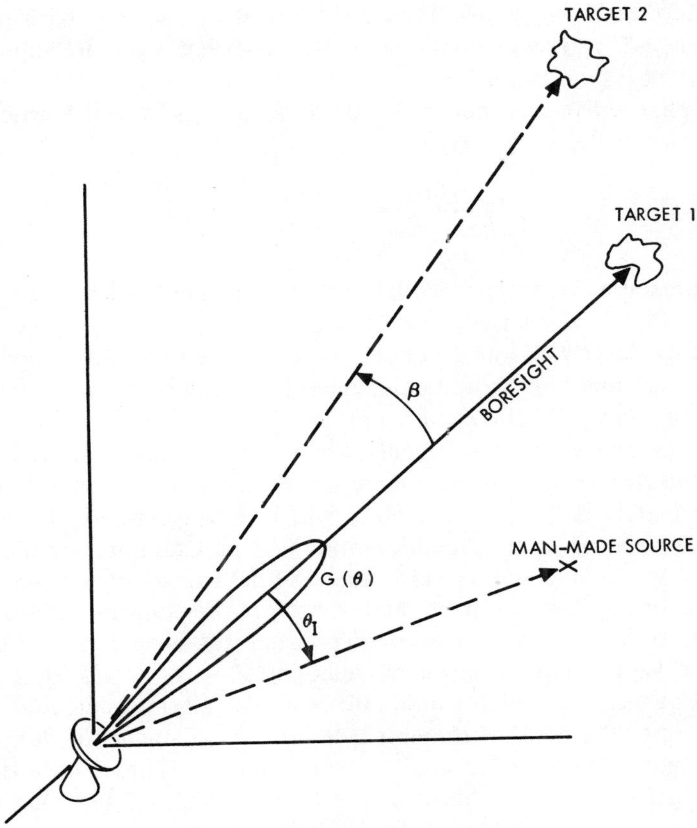

Fig. 3. Antenna coordinate system.

where $A_e(\theta)$ is the antenna effective aperture area at an angle θ, then by our definition of an interfering signal (14), the flux must be bounded by,

$$\phi_I \geq \phi_j \geq \frac{p_{smj}}{A_e(\theta_I)} = \frac{\phi_{mj} A_e}{A_e(\theta_I)}$$

Now, if the next target direction is sought by moving the antenna pointing direction by an angle β in an arbitrary direction, the minimum antenna gain or, correspondingly, the minimum effective area that might be seen at the (fixed) angular location of the interfering source will be $A_e(\theta_I + \beta)$. Therefore, for the source to remain visible at this new pointing location, we must have,

$$\phi_j \geq \frac{p_{smj}}{A_e(\theta_I + \beta)} = \frac{\phi_{mj} A_e}{A_e(\theta_I + \beta)} \geq \frac{\phi_{mj} A_e}{A_e(\theta_I)}$$

and we can define,

$$\phi_I(\theta;\beta) = \frac{\phi_{mj} A_e}{A_e(\theta_I + \beta)} = \frac{\phi_{mj} G}{G(\theta + \beta)}. \tag{15}$$

I.e. any interferer lying within the cone defined by angle θ, with incident flux greater than $\phi_I(\theta;\beta)$, will continue to be observed with an antenna angular motion of β or less.

To visualize what this means, let us look at Figs. 4 and 5 where we have plotted the parameter

$$\frac{\phi_I(\theta;\beta)}{\phi_{mj}} = \frac{G}{G(\theta+\beta)}$$

for two representative antennas. Figure 4 is an example of a modest gain horn antenna and Fig. 5, of a large paraboloid.

In order to classify all interference sources by their continued presence in the data despite antenna motion and independent of antenna pointing, thus minimizing ambiguity, we would choose $\theta = 180°$. Such a choice will also give the largest dynamic range of interference signals which may require a second look. More and more interferers will be encompassed rather than eliminated. Only those with flux greater than $\phi_{I\max} = \phi_m G/G_{\min}$ will be always ruled out, having been noted as false alarms upon antenna motion. As an example, for the antenna in Fig. 3, this would include a more than 80-*dB* range of signals that might disappear upon antenna motion and therefore be falsely regarded as ETI possibilities. However, the skirts of directional antennas fall off very rapidly. What would be the consequence of selecting $\phi_I < \phi_{I_{\max}}$? If, for example, we chose ϕ_I to be that flux which would induce a false alarm if detected at the angle corresponding to the peak of the first side lobe, then signals of flux $\phi > \phi_I$ will generally remain visible under small antenna motions. This is because the slope of the gain pattern beyond the first side lobe is small enough that motions of

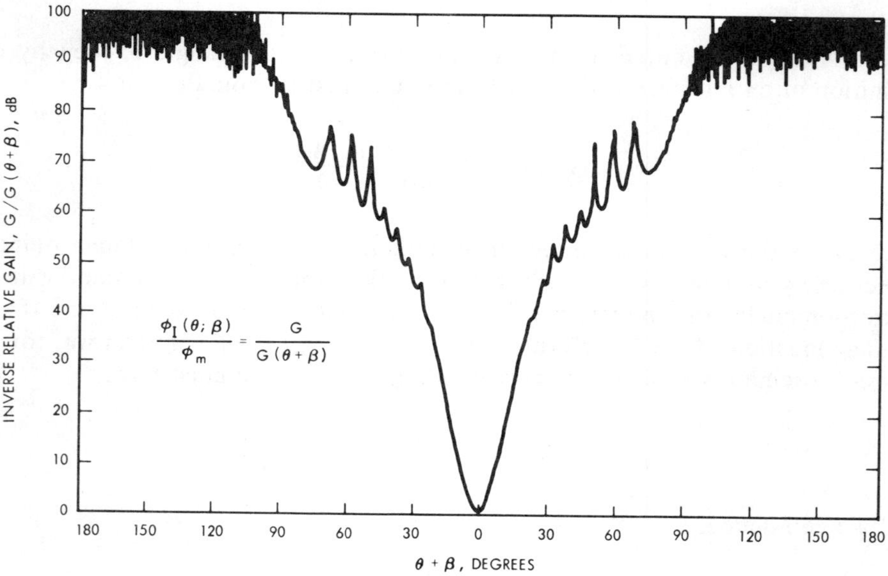

Fig. 4. Modern corrugated conical horn antenna.

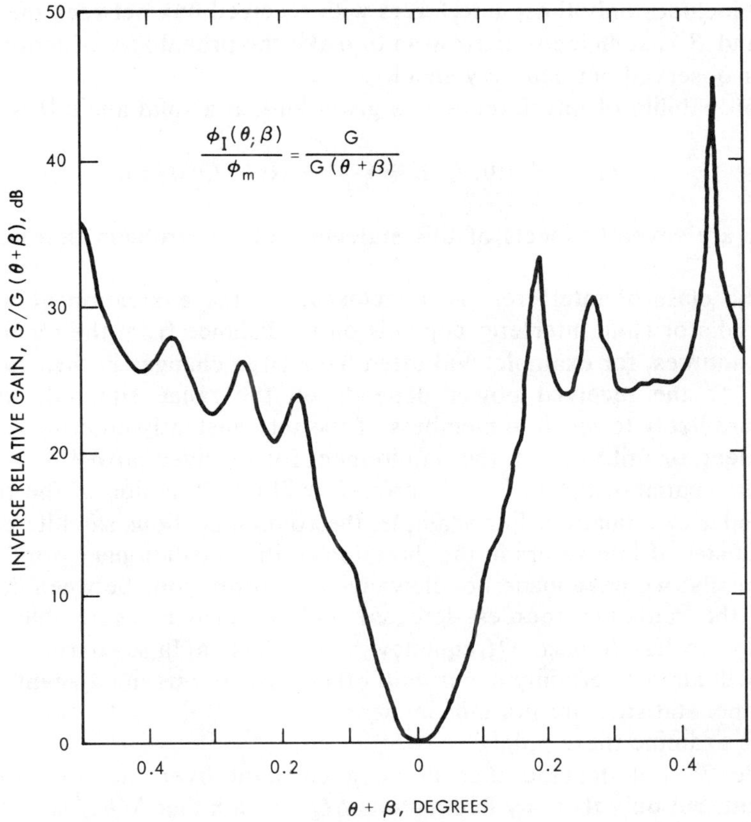

Fig. 5. Typical large paraboloid.

$\beta < \theta$ will generally result in a signal detectable at θ, remaining detectable. Thus we can say that signals with flux less than ϕ_I will require a second look if they lie within θ. Those that have $\phi > \phi_I$ will remain visible upon repointing. This then somewhat limits the number of important interfering sources.

Thus, a good strategy for determining the origin of a detected signal is to perform adjacent observations in order to eliminate the maximum number of interferers without repeated observation. This is particularly applicable for a sky survey where such a position is indeed the next target.

What will be the characteristics of the probability distribution of an interferer appearing in a single spectral bin? By the restrictions we have placed on the problem thus far, it appears likely that bothersome sources of interference will be confined to those from either airplanes or satellites within the line of sight of the antenna. By analogy with the problem of stellar distributions, we can say that the probability of seeing exactly q interferers in a solid angle Ω and a bandwidth B, given a density of Q interferers per steradian per B Hz in the celestial sphere, is represented by the Poisson distribution:

$$p(q; Q\Omega/4\pi) = [(Q\Omega/4\pi)^q/q!]\exp(-Q\Omega/4\pi)$$

where Q includes only those interferers with received flux between the limits ϕ_m and ϕ_I, and B is sufficiently narrow as to make the probability of more than one interferer observed per cell very small.

The probability of interference in a given look at a solid angle Ω is then:

$$p_I = 1 - p(0;\ Q\Omega/4\pi) = 1 - \exp(-Q\Omega/4\pi). \qquad (16)$$

There are several aspects of this statement of the problem that bear closer scrutiny.

(1) The class of interferers is not closed, i.e. the evaluation of whether a specific radiator is an interferer depends on its distance from the receiving site. Airplane sources, for example, will often have large changes in their range from the site. As the received power depends on the range squared, such large changes are likely to result in members of the set constantly entering and exiting as they meet, or fail to meet, the requirement for received power.

(2) The spatial distribution of interferers will be a function of the frequency regime under examination. For example, the communications satellite bands will show a cluster of interferers in the direction of the geostationary orbit.

(3) Finally, we have made no allowance for correlations between detections. Most of the radiating sources detected will produce a signal that occupies several to several thousand frequency bins. Thus, a false alarm in a single channel will almost certainly occur with other false alarms in adjacent channels. Interchannel statistics are not independent.

Let us examine these points:

(1) We do not demand that Q be a constant over the duration of the experiment, but only that any fluctuations ΔQ be such that $\Delta Q/Q$ is small during an observation; i.e., we require that the density of sources remain approximately constant. Since a full-frequency band NB will require a search of at most several months, the interferer environment should be relatively constant and this criterion substantially satisfied.

(2) While certain man-made microwave sources will show direction preferences, this is only true for small regions of the entire plausible frequency range. Therefore, in this analysis we will ignore this possibility. If extensive work is to be performed in such bands, the concentration of man-made radiation in particular directions will possibly exclude examination of certain sky regions.

(3) In constructing (12) above, the assumption of channel independence was made, but, indeed, for interference this assumption may be invalid. However, let us look into the definition of p_I a little more closely; p_I is the probability that in a single channel a detection is made that is caused by a radio signal from a man-made contrivance. Now let us look at N channels, where we ask, "What is the probability that in none of these channels does an interfering signal appear?" If the probability of a signal appearing in channel j is p_I, then the probability of a signal appearing in channel $j+1$ is greater than p_I, given that a signal is detected in channel j. Thus, in an actual experiment we will observe a certain "burstiness" in the distribution of alarmed channels. The extent of this clustering will depend on the individual channel bandwidths as well as the spectral characteristics of the interferer.

Thus, when using the p_I as defined, we will obtain a value of P_f that places an upper bound on the true value of P_f, that which would take the burstiness into account. That is, the model used for this simplified discussion will actually present an upper bound on the probability of false alarm in the bandwidth NB for a given value of p_I. Therefore, by using the expression of (12), we are to some degree overestimating the extent of the problem. In the absence of experimental measurements, this is an appropriate approach to providing a sufficient performance margin for the system design.

V. Example: Implications of false alarms for a constant beamwidth sky survey (Case III)

Let us examine the system implications for $P_{fi} = P_f$, independent of frequency. Thus,

$$Q_i = Q = S/2\pi N \tag{17}$$

where S is the number of sources in a bandwidth NB and within an observed hemisphere (2π steradians). Substituting (13), (16), and (17) in (12), we have:

$$P_f = 1 - [\exp(-S\Omega/8\pi^2 N) - (1/2)\text{erfc}(\alpha/\sqrt{2})]^N. \tag{18}$$

Let us see what is involved in making the probability of false alarm palatable. Earlier, in (9), we defined system degradation by the expression:

$$D(dB) = 5\log(1 + P_f).$$

If we choose an allowable degradation of 5% ($P_f = 0.05$) corresponding to a loss in sensitivity of 0.1 dB, then $p_f = 5 \times 10^{-8}$. Let us allocate half of this quantity to noise fluctuations, then $p_N = 2.5 \times 10^{-8}$ and $\alpha = 5.5$. In other words, this performance can be achieved by setting a threshold at 5.5 times the noise standard deviation. (At $\alpha = 6$, $p_N = 10^{-9}$, the curve is very steep.) Now let $p_I = 2.5 \times 10^{-8}$, then $S\Omega/N = 1.9 \times 10^{-6}$ or, noting that Ω can be expressed in terms of the beamwidth of an antenna as:

$$\Omega = 2\pi(1 - \cos\theta)$$

where θ is the half-angle of the beam, then we can plot a family of curves of degradation vs. θ for various values of α, S, and N. Figure 6 presents such a plot for $N = 10^6$ and $\alpha = 5.5$. In this example, for $D = 0.1\ dB$ and $\theta = 7.5°$ (15° antenna beamwidth), $S = 35$ sources in the band NB over the hemisphere.

VI. Experiment protocol

The above considerations suggest an experiment protocol that may be regarded as an algorithm by which to identify data of interest for further investigation (more than two looks) in the conduct of a search. In what follows, we restate our assumptions and lay out the implications in terms of a protocol.

A signal intended for detection is likely to have a simple structure. We have made the assumption that the signal we seek is basically a narrow-band trans-

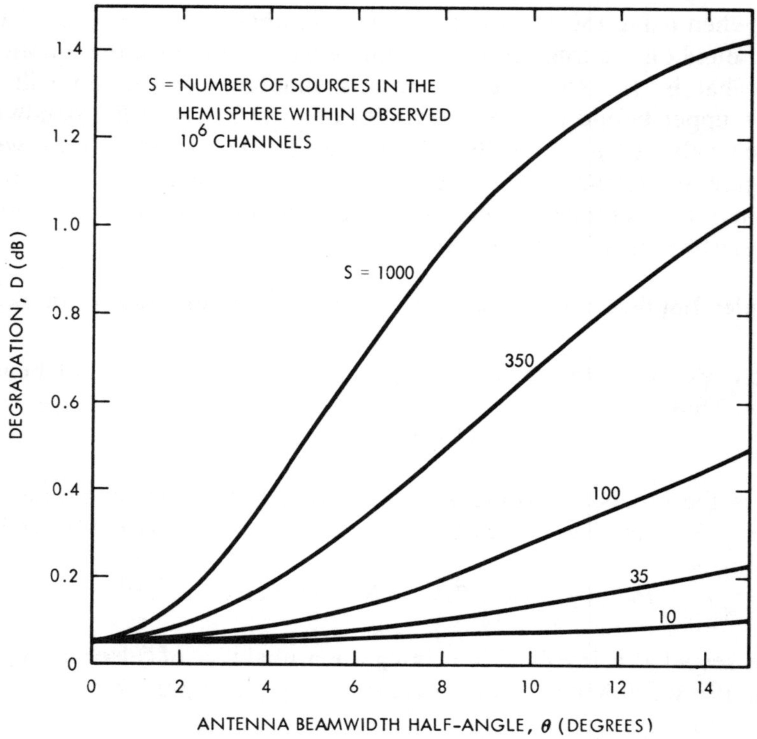

Fig. 6. Example of search sensitivity degradation under interference conditions; $\alpha = 5.5$, $N = 10^6$.

mission with a modicum of modulation. Narrow-band transmissions are profuse in the terrestrial environment and may be generated by any number of terrestrial applications. Typically, such signals will enter the receiver through the side lobes of the antenna. The probability of their being present in the main beam is small. For the elevations at which we will operate, signals in the main beam will be mostly due to Earth satellites or aircraft. Such signals will be transient. Thus, the first detection criterion is that *the signal's presence must be repeatable upon several observations.*

Strong terrestrial signals entering through the antenna side lobes will obey this criterion. However, a redirecting of the antenna pointing will leave their spectrum intact. Thus, the second detection criterion is that *the signal source must be fixed in celestial direction.* A further qualification is that *the Earth's diurnal Doppler signature must be present on the signal.* This qualification need only be tested for very suggestive signals using very narrow-band examination after their discovery in the normal detection mode.

Astrophysically generated signals may meet these criteria. So far as we know, however, signals of natural origin are inherently wide band; the presence of a signal in three or more adjacent channels would suggest natural origin. Thus, another criterion is that *the signals must be isolated in frequency.* Care must be taken in the application of this criterion to preclude omission of signals that coincidently are adjacent to signals of more prosaic origin.

Finally, if there exist signals of natural origin that are indeed narrow band, the final test will be to *ascertain whether data modulation is present.* If it is, we presume detection and call in the decryptors. If it is not, we presume natural origin and call in the astrophysicists to consider how such a thing can exist.

These criteria must be defined in detail to avoid an intolerable false-alarm problem. They must be implemented in an automated alarming system for unattended operation of the search system.

It is not unlikely that the single largest detection problem will result from intentional spoofing of the search system. We must be very sure that the first announcement of a detection is made only once. By designing the system to be immune to hoaxes at an early stage, we can significantly improve the criteria for an assured detection. The characteristic of an actual interstellar signal that is least subject to intentional mocking is the source location. By maintaining a criterion that the source must lie in the main beam of the antenna and be fixed on the celestial sphere, we can preclude all but the most sophisticated of practical jokers. Spoofing the signature of main beam detection would require access to the actual tracking status of the antenna to permit the switching of the mock signal on and off at the appropriate times. Besides efforts at securing the system against such intrusion, there is one other action that can virtually assure that this form of hoax will be detected. This leads to the final criterion of this set. *The suspect signal must be detectable by another observatory which is isolated in RF environment from the locale of the main search system.* Before calling for such assistance, we will have performed all possible verifications of detection at the observation site and will be nearly ready to make an announcement of success.

With this preface, Table 1 presents guidelines for signal extraction algorithm development. The characteristics of the two types of signals considered can be examined with two processing categories: comparisons with stored data; and real-time decisions. In Table 1, the former characteristics are indicated by *S*, the latter by *R*.

The criteria for determining detection, in the order of their application, are as follows:

(1) Constant celestial direction.
(2) Narrow-band signal.
(3) Repeatable in further observations.
(4) Appropriate Doppler signature.
(5) Data modulation present.
(6) Repeatable from other observation sites.

VII. Conclusions

This paper has explored some of the aspects of signal extraction in a microwave search for evidence of extraterrestrial intelligence. The major influences upon conducting an efficient program are the effects of the noise

Table 1. Guidelines for signal extraction algorithm development

Signal Characteristic	Decision Point for: Declaring Possible ETI Detection	Discarding Man-Made Signals
Location	Fixed on the celestial sphere (S)	Not fixed on the celestial sphere (S) (even deep-space vehicles - most of which signals can be specified *a priori* - will vary their celestial direction, albeit slowly; slowness may be a problem for broad-beam detections)
Frequency	Fixed in frequency or slowly drifting once terrestrially related Doppler has been removed (S)	Not fixed in frequency once terrestrially related Doppler is removed (S)
Modulation	Narrow band, i.e., no more than two adjacent frequency bins (R); continuously present, with allowance for detecting receiver switches due to polarization modulation (S)	Not continuously present (S) and/or occupying more than two adjacent channels (R)
Power	Exceeds threshold in a single channel, or upon summing the outputs of two adjacent channels (R), or upon summing data from adjacent locations (S)	Observed despite repointing of antenna (S) (applies to relatively strong signals)

(R) indicates decision can be made in real time on the basis of data at hand.
(S) indicates decision requires comparison with stored data.

background, both natural and man-made, upon the signal extraction process. We have shown that the key to an efficient search is the prompt elimination of false alarms. An experiment protocol is suggested which eliminates spurious signals primarily through experiment procedural techniques involving antenna repointing, delayed repeated observations, where necessary, and storage of particular historical parameters for suspect signals. This protocol is suitable for implementation in an automated search system and results in an effective approach to signal extraction.

Acknowledgements—I should like to express my gratitude to E. C. Posner, S. Gulkis, and M. F. Easterling for their helpful and beneficial comments, corrections, and encouragement and to C. Walde, A. Nelson, R. Chandlee, and A. Sedor for their unflagging efforts in the production of this paper.

References

Janssen M. A. and Gulkis S. (1977) Parameterizing an All-Sky, All-Frequency Microwave Search for Narrow-Band Radiation of Extraterrestrial Origin. Paper Presented at SETI Workshop, Jet Propulsion Laboratory, June 29–30, 1977.

Oliver B. M. and Billingham J. (1973) *Project Cyclops: A Design Study of a System for Detecting Extraterrestrial Intelligent Life*, pp. 136–140. NASA CR-114445, Revised Edition.

Acta Astronautica. Vol. 6, pp. 163–173. Pergamon Press 1979. Printed in Great Britain

Two systems analyses of SETI

ROBERT E. MACHOL

Graduate School of Management, Northwestern University, Evanston, IL 60201, U.S.A.

and

SETI Program Office, NASA Ames Research Center, Moffett Field, CA 94035, U.S.A.

Abstract—We assume a signal from an extraterrestrial source exists in the form of coherent electromagnetic radiation. We must search over a multi-dimensional parameter space: polarization, modulation, flux level, direction, frequency, and time. Searching this space will take a hopelessly long time unless its volume can be reduced by examining many points simultaneously and by eliminating some regions of the space as being unlikely. A systems analysis of this problem has demonstrated that for the next few years the greatest benefit/cost ratios will arise from development of better data processors in the form of Fourier transform devices. The payoff is linear at least up to time-bandwidth product (number of channels) of 10^9, far beyond the present state of the art. Receiver noise-figure reduction is also cost-effective. Very large antennas, which are expensive, may ultimately be necessary.

1. Introduction

We ignore here a number of questions relevant to the search for extraterrestrial intelligence, not because they are unimportant but because we are attacking a particular narrow question. Thus we ignore all discussion of whether and where life might exist outside the solar system, any question about the nature of such beings, how we might decode messages, how (or in fact whether) we should respond to any signals which we might receive, the utility of probes, etc. We are interested here solely in the attempt to detect one single signal of intelligent origin. We assume that this signal exists and, specifically, that it exists in the form of electromagnetic radiation in the microwave region (other types of radiation and other electromagnetic frequencies are here ignored for reasons presented by Oliver (1976)). There may be many such signals, but our objective is only to detect one and to receive it over a sufficient time and with sufficient signal/noise ratio that we may, with high confidence, identify it as being an artifact of intelligent rather than natural origin.

Whether such electromagnetic signals are designed for attracting our attention or for the internal purposes of the other civilization is an important distinction which will affect much of what follows. In the former case, we call the signal a "beacon" and assume that its parameters are chosen in such a way as to make its reception (and decoding) probable. In this case, it is necessary to guess what "they" think is an appropriate set of parameters for finding a place for a "meeting." It is hard to avoid being chauvinistic in such analyses, and one must be clever (but not too clever) in deciding what they think we think and what they think we should think they think, and... In the latter case, when we are eavesdropping on a signal designed for their own purposes, the situation may be very different. On the one hand, it is clearly impossible to predict what kinds of signals might be used by another civilization; on the other hand, if they are using electromagnetic radiation, they are probably using many such signals and we need only find one of them. By analogy, if a civilization at some nearby star had a sensitive radiotelescope trained on Earth, they would discover a large number of signals of widely varying form spread all over the microwave spectrum.

An interesting conclusion of this system analysis is that a number of terms which appear to be clear and unambiguous are actually vague in the context of this problem. As will be seen below, such apparently well-defined terms as flux level and bandwidth can be given several different interpretations. The exercise of specifying exactly what one means by them is quite illuminating.

2. Pattern recognition

Pattern recognition is another serious problem which will not be dealt with in great detail here. Any detection system which we build will receive many spurious signals, both natural signals from the sky and man-made signals in the form of RFI. These may come in either through the main beam or through side lobes of the antenna, and they may appear in the output of the detection system in unexpected ways because of the nature of the filters and data-processing devices which are a part of the detection system.

We have not the slightest idea of the form of the signal which we are seeking. It may be narrow-band CW, broadband CW, pulsed, etc. The modulation, if any, may be in any of the usual amplitude- or frequency-modulation forms, or it might be some periodic shift of some such parameter as polarization.

As indicated below, it will be necessary to examine a large number of "channels" (i.e. points in the parameter space) simultaneously, and this will require some sort of pattern-recognition subsystem. Little thought has been given thus far to this subsystem, but it will be a difficult problem to solve and a difficult subsystem to implement.

In the remainder of this document, it will be assumed that detection and pattern recognition take place in real time. The protocol would then be that if a suspicious artifact were detected, that particular channel would immediately be reexamined. For many reasons, it would be desirable to maintain for at least a few minutes a hard copy of the raw IF of all incoming signals, but this is beyond the present state of the art of recording. If such improvements in recording technology should become available, some of the tradeoffs indicated below would be modified.

3. Parameter space

Within the assumptions stated above, there are still at least six variables which yield uncertainty as to the exact signal that should be searched for: (1) polarization, (2) modulation (or signal form), (3) flux level, (4) direction (i.e. ϕ and θ, or right ascension and declination, or latitude and longitude), (5) frequency, including the complicating concepts of bandwidth and frequency drift rate, and (6) time. As will be indicated, these factors lead to a multidimensional parameter space in which something in excess of 10^{20} separate points need to be searched. Obviously, sequential search of these points is hopeless, and if the search is to be completed in a reasonable time and with some hope of success, the volume of this space must be reduced in two ways: by looking at many points simultaneously and by eliminating some volumes of the space as being unlikely or unrewarding. Each of the six is discussed in a separate section below.

3.1 *Polarization*

Any polarization—linear, circular or elliptical—is possible. We believe that a beacon is most likely to be circularly polarized, but we may not receive a beacon. If we receive simultaneously in any two orthogonal polarizations, then we can detect a signal of any polarization with a loss of at most 3 db. The more simultaneous separate polarization channels that are used, the lower the expected loss on a signal of randomly chosen polarization; but it is a curious fact that the maximum loss of 3 db remains the same from two to five channels. If six different channels are used, then the maximum loss becomes 1 db. Because of the expense of building additional channels, it now appears probable that the optimal receiving system will have two separate channels: one receiving in right-circular and one in left-circular polarization.

3.2 *Modulation*

The variety of wave forms which might appear is enormous. The extremes are represented by a narrow-band CW signal and an extremely short pulse; these are represented on a power-spectrum chart (frequency versus time) as an almost vertical line in the first case ("almost" because it might be drifting slowly in frequency) and an almost horizontal line in the second. By searching only for sine waves, we will detect any possible coherent signal with perhaps some loss of S/N (signal/noise ratio). Not knowing the form of the signal for which we are seeking, we must use square-law detectors; if we knew the exact form and phase of the signal, we could use a coherent detector and pick up 3 db in S/N. However, this assertion that we will lose at most 3 db by searching for coherent sine waves assumes that the signal is essentially CW with a known frequency drift rate and known bandwidth, or that it is pulsed with a known pulse length and pulse repetition frequency. Since none of these things are known, the loss (compared to an ideal coherent detector) may be considerably greater than 3 db. It appears that the optimal strategy, at least in the initial stages of the search, will be to concentrate on search for narrow-band CW signals. Bandwidth and frequency drift rates are discussed below.

3.3 *Flux level*

Flux level is presumably measured in either W/m^2 or $W/m^2/Hz$, and either at the receiver or at the transmitter. We shall measure it at the receiver, and for the transmitter we shall use the parameter EIRP (equivalent isotropic radiated power), which is equal to the transmitted power multiplied by the antenna gain. If the transmitter should be pulsed (either because the generator itself is pulsed or because there is a rotating fan beam), then there is a difference between peak power and average power, but we shall assume CW signals and assume therefore that the two are the same.

As a surrogate for flux level, we shall speak only of S/N. This is affected by the signal itself as it reaches Earth, and by three parameters of the receiving system—namely, effective antenna receiving area, receiver bandwidth, and receiver noise temperature. In addition, S/N may be affected by parameters of the detector, especially integration time. And it is affected by all the other aspects discussed in other subsections; for example, a mismatch in polarization may result in a loss in S/N of as much as 3 db.

In some sense, the flux level is affected by postulation; that is, we can design a receiving system and calculate the minimum flux level which will give a detectable S/N (presumably in the neighborhood of unity), and then decide whether it is worthwhile pointing such a receiver in a given direction. If we point it at a star 1000 light years away, we must postulate an EIRP 20 db higher than that which we would postulate if we would point it at a star 100 light years away.

For the remainder of this paper, we will assume an arbitrary upper limit on range of 1000 light years. This is dictated partly by the fact that two-way communication begins to become meaningless if the range is larger than this, and partly by the fact that with the largest system yet considered (the Cyclops array of 1000 to 2000 100-m dishes and a very sensitive receiver), a nominal signal of EIRP 10^9 W could be barely detected at this distance, if confined to a very narrow bandwidth. Furthermore, within 1000 light years there are 1 or 2 million promising stars (F, G and K), and we may have to search this many before we have an acceptable probability of successs.

If we are making a whole-sky search, presumably the integration time will be one reciprocal bandwidth. Scanning rates and the like must be adjusted to make this work out. On the other hand, if we are conducting a selected-target search, then it becomes advantageous to rest on a given target for as long as possible with S/N increasing (within some limits) as the square root of the observation time. This is discussed further below under time.

The effective receiving area of an antenna is, of course, equal to its physical area times its efficiency. The most obvious method of increasing S/N is by increasing the antenna area. This is a "brute force" solution, and while this may eventually become necessary, it appears to be very expensive. As indicated below, the principal conclusion of this systems analysis is that greater benefit/cost ratios can be accrued in the next few years by methods other than building very large antennas. It may, however, be worthwhile to attempt to develop new methods of increasing the efficiency of antennas. At the present time, such efficiencies are typically 0.75 to 0.8 at most. Since the payoff is linear with this efficiency, it may well pay to try to bring the efficiency up to 0.9 before constructing large new antennas. However, it appears that for the next few years our search will be made with existing antennas.

A quite different view appears on receiver noise temperature. Again, the payoff is linear with this temperature. In 1971, when the Cyclops study was done (Oliver and Billingham, 1973), it was assumed that a system noise temperature of 20K was the lowest that could be achieved. Now 10K appears within the state of the art, and it would surely be worthwhile to cut this still further. The ultimate theoretical limit is the 3K temperature of the sky background, and as one approaches this, further improvement becomes exceedingly difficult and expensive. It appears that, for interim search efforts over the new few years, we should be constructing receivers and antenna feeds that will permit a system noise temperature of 10K or perhaps less, and use them with existing antennas.

Finally, there is the significant question of receiver bandwidth. This is a complicated issue discussed below under Frequency.

3.4 *Direction*

As indicated, S/N is linear with the antenna area. But as the antenna becomes bigger, the antenna beam becomes correspondingly smaller. Here the distinction between selected-target search and area search is critical. In selected-target search, which means pointing the beam successively at stars which seem likely sources, the narrowness of the antenna beam has no effect except on the possible stability of the system (i.e. for extremely narrow beams there might be some question as to whether the beam intersects the target star, but this would appear to be a minor consideration). Note that the appropriate catalog of stars necessary for selected-target search does not exist, but for the purpose of this discussion we assume that it will be constructed.

It is well known that the brightest stars in the sky are not the closest stars, and it may be, analogously, that the strongest electromagnetic signals are not coming from the closest stars. By extension of this argument, it may be desirable to search nearby galaxies, where a single antenna beam can cover a very large number of stars. However, it should be noted that if the antenna is large, a galaxy such as the one in Andromeda will require a large number of pointing directions to be completely searched.

If area search is undertaken—that is, if we wish to search the entire sky—then some compromise is necessary between S/N and the time required for the search. An antenna as large as the largest presently existing one (the 300-m dish at Arecibo) requires more than 10^7 pointing directions to search the entire sky at frequencies of about 1.5 GHz, which is the nominal frequency assumed in this discussion. At 6 GHz, however, the number goes up almost to 10^9, and since much larger antennas are contemplated, it can be seen that the number of pointing directions can become quite large.

The contemplated large antennas that might ultimately be built might be in the form of a phased array. In particular, the Cyclops study suggested an array of 1000 to 2000 dishes each 100 m in diameter and separated by 300 m (on center). Such an array is pointed in a given direction electronically, that is, by appropriately phasing the returns from the individual dishes before combining them. By using two separate data processors, one can simultaneously point the beam in two different directions and examine two different targets, provided both are within the beamwidth of a single dish. If one has n dishes in the array, then by the simultaneous use of n parallel data processors it is possible to examine the entire sky within the beamwidth of any single dish. Of course, the antenna gain pattern is not quite uniform across this wide a beam, and there can be losses up to 3 db or more, depending on exactly how the successive positions of the beam are tesselated.

Whether we are using an area or a selected-target search, a single antenna on the surface of Earth cannot cover all declinations. It may be a reasonable strategy to locate an antenna in the southwestern part of the United States where it can cover about 2/3 of the sky; or it may be desirable to have separate antennas for covering the northern and southern hemispheres; or it may be desirable to put an antenna on the Moon or in space where a single antenna can cover the entire sky.

It should be noted that, by appropriate phasing of such an array, it is possible to broaden the beam and thus complete the search of the entire sky in a shorter period of time, but at low S/N.

3.5 *Frequency*

We omit here discussion of the appropriate frequency band for the search, which has been covered by Dr. Oliver, and assume that we will search only the so-called water hole at about 1.42 to 1.66 GHz. If we should wish to search a frequency band 10 times this wide, it will be necessary to multiply things by 10 in an obvious way. There remain the significant questions of bin width and drift rate.

3.5.1 *Frequency bin*, The possibility of examining many individual frequencies simultaneously over a broad band of frequencies should always have been obvious since, for example, a spectrograph in the optical region does just that. However, in the microwave region, it was not obvious until the Cyclops study showed that it could be done and conceptually demonstrated the hardware that would do it. This hardware is not unlike that developed in the early 1960's for synthesizing the antenna of a side-looking radar. It basically performs a spectral analysis by taking the Fourier transform of the raw signals and then identifying any spikes in this frequency-domain representation (power spectrum) which must represent sine waves in the input signals.

Both analog and digital techniques are available to perform the Fourier transformation. The analog techniques are most advanced at the moment, especially optical techniques which use lenses and photographic films. Digital techniques involving the FFT (fast Fourier transform) are not yet so well developed, but it seems inevitable that they will soon be more powerful and less expensive.

The fundamental measure of effectiveness for all such devices is the time-bandwidth product, which is the ratio of total bandwidth searched (here called the "frequency band") to the bandwidth of the individual frequency (or "frequency bin"); it is thus equal to the number of "channels" the system is simultaneously capable of analyzing. The denominator of this ratio is not quite the same as the resolution capability. "Resolution" refers to the ability to separate two signals that are close together. More specifically, it is the ability to distinguish between the signals in Figs. 1 and 2, where the amplitude ratio a/b is arbitrarily specified and the "resolution" is defined by the distance x in the dimension of interest (in this case, frequency). If Fig. 2 represents the transmission characteristic of our receiver considered as a filter, then we are interested in the total noise received when the input is wide-band white noise,

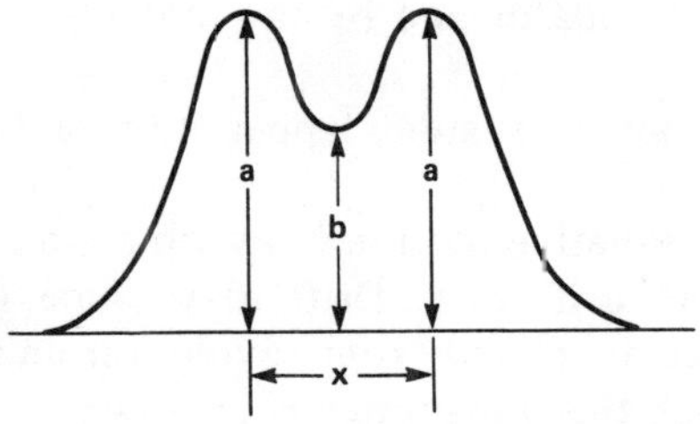

Fig. 1. Two equally strong, overlapping, but "resolved" signals.

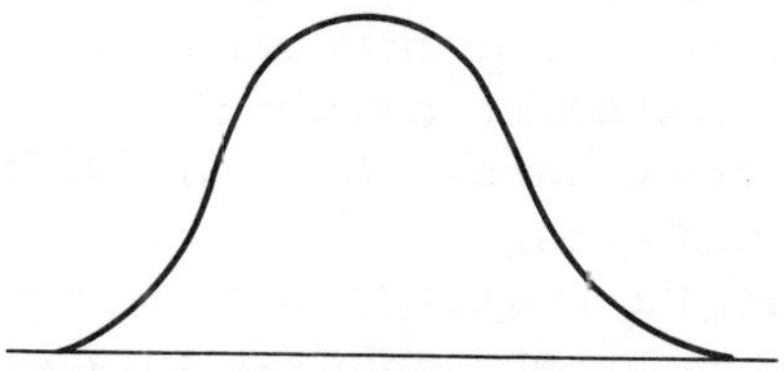

Fig. 2. Two equally strong, overlapping, but "unresolved" signals.

which is measured by the area under the curve in Figure 2 and, more specifically, by the width of a rectangle of equal area and unit height. This width may be called the "noise bandwidth" of the frequency bin. Thus the "time-bandwidth product," or TBP, is the dimensionless ratio of two bandwidths; its name derives from the fact that instead of dividing by the width of the bin we may multiply by its reciprocal, the coherent-integration time.

Since the noise power is directly proportional to the bin width, it is desirable to reduce that width as much as possible, although there is no advantage in reducing it below the bandwidth of the transmission. A beacon would presumably have the smallest bandwidth consistent with its Doppler drift rate; if the beacon were beamed and Doppler-compensated, that bandwidth might be smaller than our technological capability to build narrow-band receivers. Bandwidths below 1 mHz would probably not be of interest even if we could build them, partly because of the lack of modulation, but mostly because it requires a time of at least 1/B second for a filter of bandwidth B Hz to settle down, which would mean excessive search time.

For eavesdropping one must guess, but there is good reason to believe that bandwidths far below those of usual communication channels might be useful. Typical broadcast signals on Earth, with bandwidths measured in kHz or MHz, have a large fraction of their power in the carrier, which is often stable to a fraction of Hz (theoretically, of course, a pure, stable, unmodulated carrier has a bandwidth of zero).

A data processor with, say, $TBP = 10^6$ could be used to search simultaneously, 1 million 1-Hz bins over a 1-MHz band, or 1 million 10-Hz bins over a 10-MHz band, or 500,000 10-Hz bins in each of two orthogonal polarizations over a 5-MHz band, etc. The upper limit on the width of the band is set by the maximum speed of computation, and the lower limit on the width of the bin by the stability of the local oscillator and by the maximum Doppler drift rate that one wishes to allow for.

The TBP of existing optical systems appears to be 10^5 to 10^6, and of digital systems, 10^3 to 10^4.

3.5.2 *Frequency drift.* Relative motion between a star and our Sun will cause a frequency shift, but no drift rate. Drift may come from instability of the transmitter and/or the receiver, and from revolution and rotation of the Earth and of the planet on which the transmitter is located.

It is conceivable that the transmitter will be purposely shifted in frequency. In the famous analogy, if two people are searching for one another in a forest, it is surely better to have one stationary than to have both moving about; but the worst possible situation is to have both stationary. With this in mind, the transmitting civilization might decide to sweep slowly through a frequency band. For example, if they wished to sweep through the 242 MHz of the water hole at 1 Hz/s, it would take almost 8 years.

For planetary motion, the magnitude of the Doppler shift is greatest for revolution, while the magnitude of the drift rate is greatest for rotation. These are both sinusoidal with time, so that the maximum drift rate occurs at the time of zero Doppler shift and maximum acceleration (hence nonlinear drifts are unimportant). For rotation, this maximum drift rate occurs at the equator looking at the zenith; for other positions and directions, the drift rate is reduced by the product of the cosines of the latitude, declination, and hour angle. For Earth, the maximum is about 0.15 Hz/s at 21 cm. It goes as the second power of the angular velocity and the first power of the radius of rotation, so for an Earth-sized planet rotating at Jupiter's rate (8 hr), it would be 1.4 Hz/s, and for Jupiter (10 times the diameter), 14 Hz/s. These drift rates are linear with frequency (e.g. three times as great at 7 cm).

We can and undoubtedly should compensate for Doppler drift rates due to the rotation and revolution of Earth by appropriately drifting our own local oscillator. This is easy because we know exactly where the receiver is looking (although slightly different compensations are required at opposite ends of a wide band such as the water hole). This has the additional advantage of identifying Earth-based interference signals (coming in through side lobes) because they will now be "drifting" at a known rate. The search will usually be carried out close to the meridian, where the drift rate is maximum. If the transmitter is beaming at us in an attempt to establish contact, it can also compensate. If they have a broad-beam or omnidirectional beacon, drift can be reduced by placing transmitters at the poles or in a distant satellite; if we are eavesdropping, then we cannot expect Doppler compensation and must anticipate drifts up to some maximum.

It seems clear that bandwidths as small as 0.1 Hz are technologically achievable without great difficulty and, of course, the smaller the bandwidth, the smaller the noise power. Such a bandwidth implies coherent integration over a period of 10 sec, which is not possible unless the drift rate is less than 0.01 Hz/s, or unless compensation is made for such drift rate. If we wanted this bandwidth and this integration time, and still wanted to assume drift rates up to 1 Hz/s in either direction, it would be necessary to replicate the complete data-processing system 200 times. On the other hand, if we had a bandwidth of 1 Hz (1-sec coherent integration), then only one data processor would be required, and all possible drift rates up to 1 Hz/s could be obtained by very simple additions (a few adders), i.e. by post-detection integration on various assumed drift rates.

Shannon's law of information theory says that the information-processing capacity of any system cannot exceed its bandwidth multiplied by the factor $\log_2(1 + S/N)$. In the Cyclops study, the total integration time (1000 sec) was 100 times the coherent integration time ($1/B = 10$ sec); this extra factor of 100 can be interpreted as increasing the information-processing capacity to handle drifting signals as well as stationary ones, or increasing S/N to increase the detection probability or to reduce the false-alarm probability, or some combination of these. For $S/N = 0$ db, it is possible to have the probability of detection and the false alarm probability each of 0.5. In most schemes, it takes only a few extra decibels to raise the probability of detection to 0.99, which is adequate. Because of the large number of channels, the false alarm probability in any one channel must be reduced to very small values (10^{-12} was assumed in the Cyclops study). S/N tends to increase as the square root of the integration time.

3.6 *Time*

As indicated above, the integration time of the detector must be at least one reciprocal bandwidth. If we are engaged in selected-target search, it is probably desirable to increase this time considerably, perhaps up to 100 reciprocal bandwidths, in order to increase the probability of detection and/or decrease the false-alarm rate. Since we are talking about bin widths as small as 0.1 Hz and corresponding coherent integration times of 10 sec, we may have observation times as long as 1000 sec for each target. Clearly, if each target is examined for so long a time, there is a limit to the number of targets that can be examined within a lifetime.

The other, more difficult question regarding time concerns intermittency of signals. That is, if we have examined a particular target or a particular direction in space with some particular frequency and two polarizations and the like and have not discovered a signal, there is no guarantee that we would not discover a signal if we reexamined the same point in the parameter space subsequently. If there is a beacon, it might be omnidirectional; or it might be some sort of fan beam slowly sweeping the galaxy; or it might even be a pencil beam directed from time to time at likely targets of which we might hope to be one. If they choose one of the latter alternatives, then they might illuminate us for a few seconds every year, or for a few days every thousand years. If we are eavesdropping, then probably a signal of sufficient strength to be detected would be directional and would sweep past us because of the rotation of the transmitting planet. Again, if they had something like one of our TV stations, which concentrates most of its energy toward the horizon, the beam might sweep by in about an hour; but something like one of our high-gain military radar antennas or microwave relay links might sweep past in a matter of minutes. For all these reasons, our search strategy will undoubtedly include frequent reexamination of exceptionally promising targets; namely, the stars very similar to our own which are comparatively close by. Furthermore, as more sensitive systems become available, we will undoubtedly wish to reexamine points which have already been searched at lower sensitivity. This consideration is hard to quantify at the present time.

4. Magnitude of the problem

The justification of the earlier statement about the magnitude of the search can now be given by multiplying together the number of channels in the data processor required by all of the considerations in the preceding section. We start with two polarizations. Then, assuming a bin width of 0.1 Hz and a bandwidth (across the whole water hole) of 250 MHz, we have 2.5×10^9 different frequencies. We might then also want 200 different frequency drift rates (to pick up drifts up to 1 Hz/s), 10^6 different pointing directions for the array, and 10^3 different beams for each pointing direction (these 10^9 pointing directions come out as the same requirement if we use a single, very large antenna). We thus have a product of 10^{21} points in the parameter space, each of which must be examined for 10^3 sec. If they were examined sequentially, the search would take something over a million times longer than the age of the universe.

Obviously, we have to give somewhere. We start by widening the frequency bin, which gives us a linear reduction in S/N (only if we increase it beyond the bandwidth of the actual CW signal) and decreases the number of channels required by the cube of the bin width. This cubic factor is because the number of channels required to handle a given drift rate goes inversely as the square of the bin width, and the total number of frequency channels goes linearly. Second, we should probably be sanguine about frequency drifts and hope that they will not be high. Third, we should undoubtedly start with the selected-target search, giving us initially a few thousand and ultimately at most a few million directions in which to look, instead of 10^9. Finally, we may choose to cut our observation time.

Consider, for example, the following set of specifications: two polarizations, 1 million pointing directions (the most likely stars), bin width of 0.5 Hz yielding 5×10^8 frequency channels in the water hole, and 2-sec coherent integration time picking up all drift rates up to 0.25 Hz/s. One-hundred coherent integration times now become 200 sec, and if we had a data processor with 10^9 channels, we could complete the search in about 7 years. Allowing for repeated searches of the most promising targets, we might make it in 10 years.

Other tradeoffs are obviously possible, but the key point to note in the above suggested list of parameters is the 10^9 channels in the data processor. The appropriate data processor is a Fourier transform device; NASA Ames is now designing a device using a new FFT algorithm and new circuitry which could be completed in a year or two, and which would have 10^6 complex channels or 2×10^6 real channels. If this device is successful, we hope to be able to expand it. If we are very lucky, within 5 more years we might have a device with 10^9 channels at a total cost not greatly in excess of \$$10^7$.

We have a number of directions in which we could go at the present time, and all of these are under careful study at NASA-Ames. One is to build larger and more efficient antennas, possibly on the Moon or in space. This may eventually be necessary. Another is to build better low-noise receivers, and such work is under way, especially at JPL. A third is to build Fourier transform devices with greater time-bandwidth products. It appears that the payoff, in terms of expected number of detections of extraterrestrial intelligence (this is a technical term roughly equivalent to the probability of a detection), is linear with all three of these, that is, with effective antenna area, with the reciprocal of system noise temperature, and with time-bandwidth product of the data processor. For the near—term future-over the next 5 to 10 years—it appears that the greatest cost/benefit ratios will come from the data processors and from the low-noise receivers. Meanwhile, it is essential that we continue our studies and design efforts on larger antennas together with such ancillary topics essential for a successful search as the star-catalog studies and the pattern-recognition subsystem.

The assertion concerning payoff function gives a clue to the second systems analysis. We should like to be able to make other assertions about cost/benefit ratios applying to various aspects of the hardware configurations and search strategies which we are designing for SETI. This requires additional parameters characterizing the transmitted signal. We might generate reasonable probability distributions over kinds of planetary motions, types of signals, power, antenna gain, and the like, and see how these affect the desirability of various receiving systems, in the hope that invariance patterns will emerge showing certain hardware configurations and search strategies to be most promising in a wide variety of putative situations. This second systems analysis is barely under way.

References

Oliver B. M. and Billingham J. (1973, Revised Edition) *Project Cyclops: A Design-Study of a System for Detecting Extraterrestrial IntelligentLife*, NASA CR-114445.

Oliver B. M. (1976) *The Rationale for the Water Hole.* IAF Paper IAF-A-76-32; *Acta Astronautica* **6**, 71–79.

Acta Astronautica. Vol. 6, pp. 175–201. Pergamon Press 1979. Printed in Great Britain

An infinitely expandable space radiotelescope†

V. I. BUYAKAS, YU. I. DANILOV, G. A. DOLGOPOLOV, K. P. FEOKTISTOV, L. A. GORSHKOV, A. S. GVAMICHAVA, N. S. KARDASHEV, V. V. KLIMASHIN, V. I. KOMAROV, N. P. MELNIKOV, G. S. NARIMANOV, O. F. PRILUTSKY, A. S. PSHENNIKOV, V. G. RODIN, V. A. RUDAKOV, R. Z. SAGDEEV, A. I. SAVIN, YU. P. SEMENOV, I. S. SHKLOVSKII, A. G. SOKOLOV, G. S. TSAREVSKY, V. I. USYUKIN AND M. B. ZAKSON

Space Research Institute, Academy of Sciences, Moscow, U.S.S.R.

Abstract—Great rewards for radio astronomy lie in increasing antenna collecting areas and the distance between them. Both these possibilities, which determine sensitivity and angular resolution, can be developed essentially without limit if radiotelescopes can be built in space.

Radio interferometers, with base lines up to 10 AU, assembled from large reflectors 1 to 10 km in diameter, could operate in the 1 mm to 1 m range with a sensitivity of 3×10^{-37} W m^{-2} Hz^{-1} and an angular resolution of 1.5×10^{-10} sec of arc. These parameters are a million times better than those of modern radio astronomical systems. Such instruments could solve many urgent problems: the search for artificial radio signals and other evidence of extraterrestrial civilizations; detection and study, within the RF range, of solar-like stars and planets; and holography of astronomical sources and direct measurements of their distance. The maximum distances of objects which could be studied would be comparable with the radius of curvature of the Universe.

A possible space radiotelescope (SRT) configuration, with a modular spherical reflector and feeds providing multibeam operation, is analyzed here. An extendable reflector capable of operating at any intermediate stage is assembled from 200-m modules. These consist of a three-dimensional spar framework on which flat hexagonal subreflectors of about 4 m are mounted. The precise geometry required of such a reflector is provided by automatic adjustment of subreflector position relative to the framework and by adjustment of connections between modules. In this way, radiotelescopes for the 1 mm to 1 m range may be built with diameters up to 10 km.

The problems of system delivery and assembly in orbit are assessed.

Preliminary estimates for developing and constructing an SRT that is more than 1 km in diameter are lower than for a similar collecting area on Earth. State-of-the-art technology already allows the development and construction of a 200-m space module.

1. Introduction

THE MAIN trends of space studies that can be carried out with reusable space vehicles have recently become a subject of extensive discussion. One important objective can be the deployment in space of mirror radiotelescopes with collecting areas many times that of ground-based antennas. Such space radiotelescopes (SRT) can solve several important problems in contemporary astrophysics (Kardashev *et al.*, 1971; Kardashev *et al.*, 1973; Sheaffer, 1976; Preston *et al.*, 1976).

---

†Russian version of this paper is published in *Kosmicheskie issledovanija* **16**(5), 773, 1978 and **16**(6), 921, 1978.

A typical feature of radio astronomy is its ability to study objects in a regime where extraordinarily low fluxes are the general rule. This is possible because of the exceptionally low intensity of the background radiations in the radio-microwave range, and to the low energy of radio quanta. Hence, for a given radiant power, the best signal/noise ratios are obtainable in this spectral region. This permits one to identify and study the weakest sources of radiation, including objects with temperatures so low that they radiate only in the radio frequency range. It also explains why the most distant objects in the Universe known to us were first detected by radio astronomers.

It is natural that the development of an SRT should necessitate the solution of many complicated scientific and technological problems; however, opportunities opened up by SRT applications make this concept very promising. State-of-the-art technology makes it possible to develop a large SRT in the near future.

The advantages of space radiotelescopes are obvious for reflectors of several kilometers filled aperture, if they are capable of operating at decimeter, centimeter, and millimeter wavelengths. Such systems can be built with modular reflectors (where antennas are extended by putting together standard modules 100 to 300 m in diameter) and by automatic adjustment of reflector surfaces (Fig. 1).

The basic characteristics desired for large space radiotelescopes, and the associated problems, are discussed below.

2. Astrophysical problems and basic instrumental characteristics

The sensitivity and angular resolution of a radiotelescope are basic parameters that determine whether it can be used for particular radio astronomical studies. In a system with two identical reflectors operating as an interferometer, its sensitivity, i.e. the rms value of the threshold spectral density of a radiation flux, is determined by the well-known ratio:

$$\sigma_F = \frac{\sqrt{(2kT)}}{A\sqrt{(\Delta\nu\Delta t)}} \tag{1}$$

where k is the Boltzmann constant, T is the system noise temperature, A is the effective area of a reflector, $\Delta\nu$ is the receiver bandwidth, and Δt is the signal integration time. The angular resolution of an interferometer, its half-power lobe width, is determined by

$$\phi = \lambda/B \tag{2}$$

where λ is the operating wavelength and B is the projection of the base line of the two antennas on the wave-front plane.

At present, sensitivity can be improved considerably mainly by enlarging the effective area, A, since the noise temperature of radiometers is near that of the cosmic background, the receiver bandwidth cannot exceed its frequency, and integration time has already reached several days.

It is not easy to make the area of ground-based antennas larger, partly

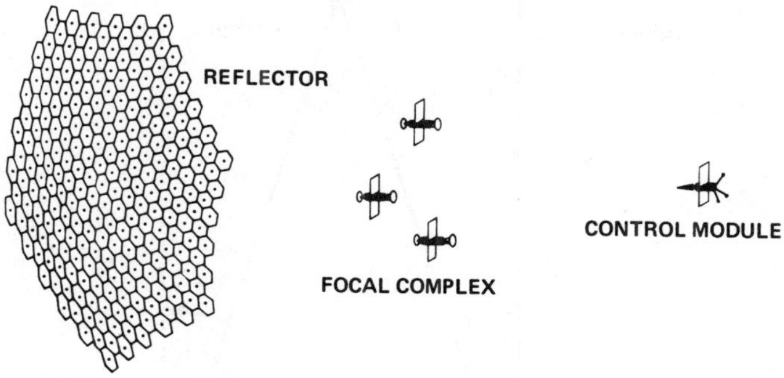

Fig. 1. Infinitely expandable radiotelescope. Primary spherical reflector is assembled from hexagonal 200-m modules, on which flat reflecting elements of about 4 m are mounted. The positions of each flat element and each module are adjusted by commands from a control module in the center of the sphere. The focal unit includes several autonomous receiving elements that provide multibeam operation.

because of the effect of atmosphere irregularities. As for angular resolution, it can be improved for a given wavelength by extending the base line, B. At present, antenna separation of ground-based interferometers has reached its limit, i.e. the separation has almost equalled the diameter of Earth. Hence the only way to enlarge parameters A and B is to place radiotelescopes in space.

Let us consider the potential of SRT's by assuming, as initial parameters, antenna diameters from 1–10 km, wavelengths from 1 mm to 1 m,† a maximum base line of 10 AU, $T = 10°K$, $\Delta\nu = 1$ GHz, $t = 1$ day, and an antenna efficiency, in terms of its geometric area, of 80%. Then the effective area of a single SRT would be 6×10^5 to 6×10^7 m^2 and its angular resolution 200 to 0.02 are sec. The threshold sensitivity of a two-element interferometer system would then be 3×10^{-35} to 3×10^{-37} W/m^2-Hz (or 3×10^{-9} to 3×10^{-11} jansky) and the angular resolution, 1.5×10^{-7} to 1.5×10^{-10} arc sec, millions of times better than is possible with ground-based antennas.

A radiotelescope with such a great sensitivity will simultaneously receive radiations from a great number of sources within the antenna beam (confusion effect). However, the interferometric regime allows an image of objects within the antenna beam to be synthesized and the above effect removed. This is why a simultaneous increase in both the antenna area and the base line is required.

Three-dimensional (holographic) images of sources observed and direct measurements of distances to them are new possibilities provided by space interferometers (Kardashev *et al.*, 1973) (see Fig. 2). Maximum distances for which such measurements can be carried out, for the given wavelength band and

†It is technically most difficult at present to provide a millimeter-range SRT. However, such systems are under consideration even for the optical range (Sky and Telescope article, Anon. 1977). As a first effort, SRT performance at millimeter wavelengths can be studied through the design and test of a single SRT module in the 100–300 m size range.

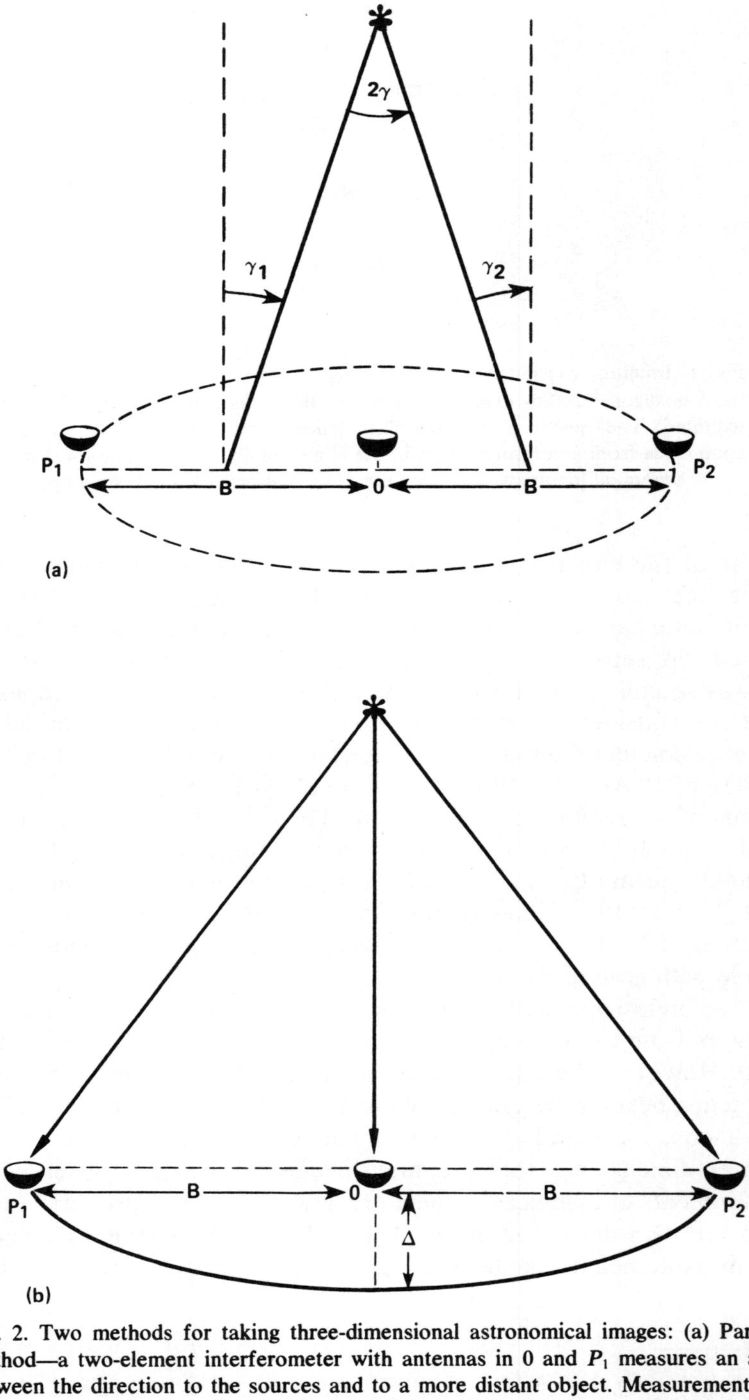

Fig. 2. Two methods for taking three-dimensional astronomical images: (a) Parallax method—a two-element interferometer with antennas in 0 and P_1 measures an angle between the direction to the sources and to a more distant object. Measurements are repeated at the opposite part of an antenna orbit in 0 and P_2; parallactic angle $2\gamma = \gamma_1 + \gamma_2 = B/R$. (b) Wavefront-curvature method—three antennas operate simultaneously measuring Δ by a phase method; distance $= R = B^2/2\Delta$.

$B = 10$ AU, are equal to the Fresnel zone radius:

$$R = \frac{2\pi B^2}{\lambda} = 1.4 \times 10^{27}\text{–}1.4 \times 10^{30}\ \text{cm}$$

$$= (1.5\text{–}1500) \times 10^9 \ \text{light years} \qquad (3)$$

which is comparable to the radius of curvature of the Universe (10 to 20×10^9 light years). All of the Universe turns out to be in the near (Fresnel) wave zone for this system! It should be emphasized again that only in space are these experiments possible. For the maximum ground base-line of 10,000 km and minimum wavelength of 1 cm (because of Earth's atmosphere), eqn (3) yields $R = 6 \times 10^{18}$ cm $\simeq 6.5$ light years as a limit, within which there are only four of the nearest stars.

Another essential advantage of space radiotelescopes is the fact that artificial and industrial radio interference can be avoided. Their level on Earth is fairly high and growing continuously. This interference can be removed by putting an SRT farther away from Earth; by the screening the reflector provides for itself when operating in directions away from Earth; and by building screens in space. According to international agreements, only narrow regions of the RF range, about 6% of the 13 octaves of the radio frequency spectrum, are allocated to ground-based radioastronomy; hence there is no radio interference there. With an SRT, that value can be increased to 100%, i.e. the entire spectrum could be used.

The development of space radiotelescopes opens up new and grand prospects for radioastronomic studies. Several examples of the most important scientific problems that can be solved with SRT systems are given below.

Detection of extraterrestrial civilizations

Projects involving the search for artificial radio signals discussed by various authors (Sagan, 1973; Oliver & Billingham, 1973; Anon., 1974) confirm the necessity of building large radiotelescopes usable over a broad wavelength band. First, observations should be carried out to search for information-bearing (artificial) radio signals and, second, for very big solid-state structures, the product of astroengineering activity.

With the help of eqn (1), the distance limit, R_1 (at which it is possible to detect radiation from a transmitter with an average power, P, in the bandwidth, $\Delta\nu$, and with a transmitting antenna gain, G) is

$$R_1 = \left(\frac{PGA}{4\pi kT}\right)^{1/2}\left(\frac{\Delta t}{2\Delta\nu}\right)^{1/4}. \qquad (4)$$

With the assumed SRT parameters ($A = 1$ to 10 km dia.) and with $P = 1$ MW, $G = 1$ (an isotropic radiator), $\Delta\nu = 100$ Hz, the distance will be $R_1 = 10\text{–}100$ light years. This distance will naturally be longer if radiation is not isotropic and is directed toward the solar system ($G > 1$). Thus, for instance, for 50 dB of transmitting antenna gain (antenna diameter = 15 cm at 1 mm, =1.5 m at 1 cm, and $= 15$ m at 10 cm), the distance at which detection is possible increases to 3–30 thousand light years, which is comparable with the dimensions of our Galaxy.

Dyson's sphere-type astroengineering structures can be discovered by their screening of the background radiation.

For a Dyson sphere around a star of luminosity, L, if eqn (1) is used, the distance limit, R_2, at which such a structure could be observed is,

$$R_2 = \frac{(2\Delta\nu\Delta t)^{1/4}}{2\lambda T_D^2}\left(\frac{AL}{\sigma}\right)^{1/2} \qquad (5)$$

where T_D is the equilibrium temperature for a Dyson sphere and σ is the Stefan–Boltzmann constant. For the assumed SRT parameters ($D = 10$ km, $\lambda = 1$ mm) and a luminosity about that of the Sun, R_2 is about 1.5×10^6 light years for $T_D = 300$°K and 1.5×10^{10} light years for $T_D = 3$°K. The former is comparable to the distance to the nearest Galaxies, the latter to the radius of curvature of the Universe.

For artificial structures to be detected by their proper radiation, the distance limit R_3 can be determined from,

$$R_3 = R_2(T_D/T)^{1/2} \qquad (6)$$

Only for $T_D \gg T$ can this result be better than eqn (5). If $T_D = 300$°K, then $R_3 = 8 \times 10^6$ light years.

Figure 3 illustrates the detection possibilities described above.

Detection of stars and planets

Estimates based on eqn (1) show the possibility of detecting and studying radio radiation from stars similar to the Sun at distances up to 10,000 light years

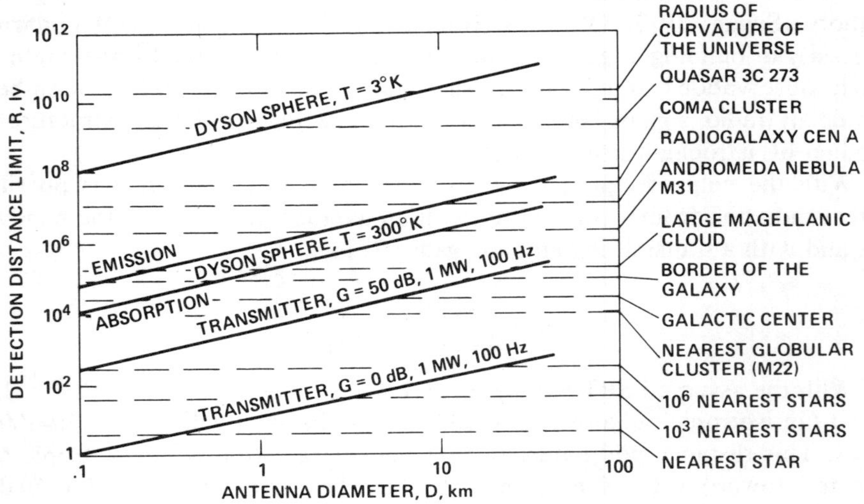

Fig. 3. Possibilities of detecting extraterrestrial civilizations with an SRT.

at $\lambda = 1$ cm and up to 100,000 light years at $\lambda = 1$ mm. Note that no radiation from a solar-type star has ever been recorded. At wavelengths of 1 cm and 1 mm, planets like Jupiter can be detected at distances of 150 to 1500 light years, and Earth-type planets at distances of 20 to 200 light years. Emphasis should be given to the fact that there are about 100 stars within a 20-light year radius centered on the Sun, many of which may have planetary systems.

Figure 4 shows the capabilities of detecting stars and planets.

Holography of the distant sources in the Universe

As mentioned above, an SRT system can be used to obtain three-dimensional images of the sources studied and to measure their distances by trigonometric methods. In fact, the whole Universe can be holographed with interferometer base lines of the order of 10 AU (see eqn (3)). During the holographic process, the region covered by a source is subdivided into individual elements, each less than *B*. A radiotelescope should be capable of recording radiation from each element. Whether such an experiment is possible will be estimated from the sensitivity of the radiotelescope (eqn (1)) and from the radiative power of well-known radioastronomic objects.

As is well known, the radio radiation of the most powerful extragalactic sources (quasars, radio galaxies) is associated with the motion of relativistic electrons in magnetic fields (synchrotron radiation). The brightness temperature of such radiation is limited by self-scattering and cannot exceed $T_{max} = 10^{12}$ K (Kellermann and Pauliny-Toth, 1969). Hence the maximum radiative flux from the source is,

$$F_{max} = \frac{2kT_{max}}{\lambda^2}\frac{\pi\alpha^2}{4}$$

where α is the angular diameter of an element of the source as seen by the

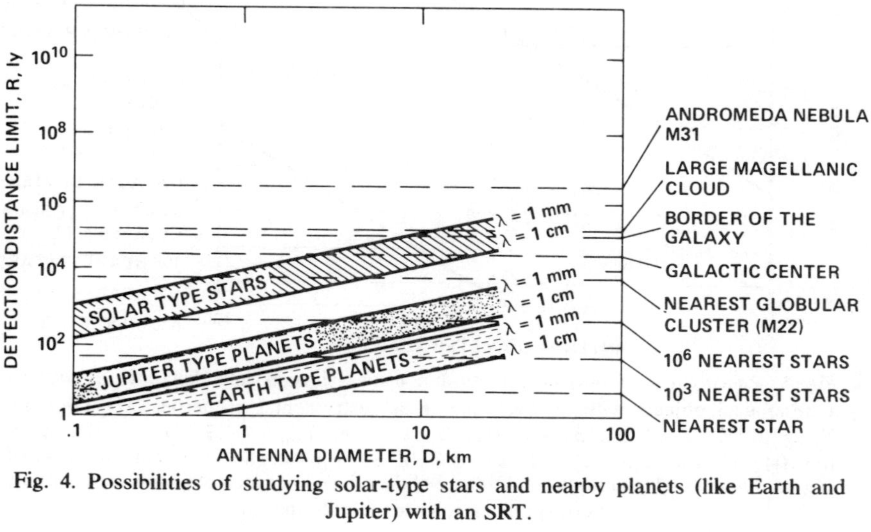

Fig. 4. Possibilities of studying solar-type stars and nearby planets (like Earth and Jupiter) with an SRT.

observer. In holography, this diameter must not exceed the beamwidth of the interferometer ($\alpha \leq \lambda/B$). Then, using eqn (3), we find $F_{max} = \pi^2 k T_{max}/\lambda R$. Comparison of F_{max} with the flux limit of eqn (1) gives the distance, R:

$$R = \frac{\pi^2 k T_{max}}{\lambda \sigma_F}. \tag{7}$$

For the assumed parameters and for $\lambda = 1$ cm, distance R is about 4×10^{12} light years, which exceeds considerably the radius of the observable Universe. Figure 5 demonstrates the capabilities of holographic methods.

3. SRT system and its geometric parameters

We now assume that a space radiotelescope should consist of a main mirror, modularly extendable, several autonomous radio receiving elements, and a controlling element on the geometric axis of the main mirror (Fig. 1). Off-axis receiving elements are there to provide multibeam reception from any direction within some angle θ relative to the axis of the primary without its reorientation. The control unit measures the figure of the primary mirror surface and the position of the receiving elements and controls the system by sending commands to adjust the surface and reorient and move the elements.

The primary mirror should be spherical to provide multibeam operation under which several distant sources can be observed simultaneously. It will also reduce the requirements with respect to the accuracy of its orientation; high pointing accuracy is provided by properly adjusting the position of the receiving

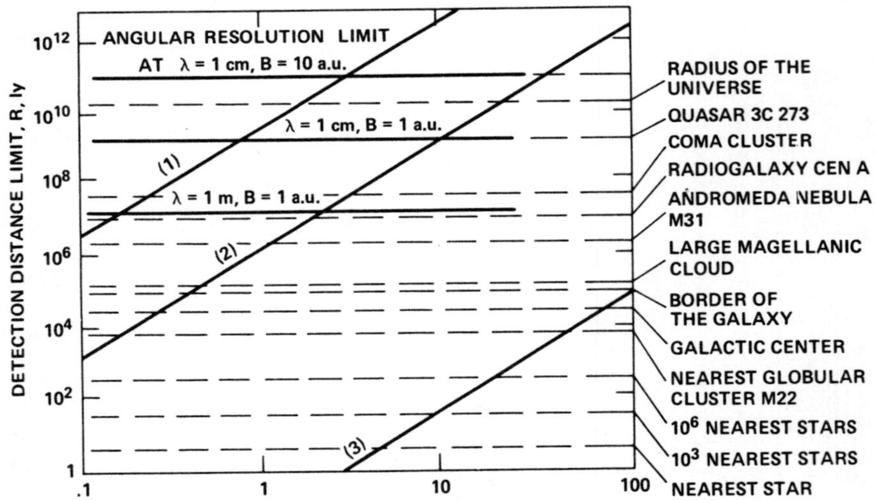

Fig. 5. Possibilities of obtaining three-dimensional images and distances of various astronomical objects with an SRT and these instrumental properties: (1) $\lambda = 1$ cm, $T_{max} = 10^{12}$ K, $T = 10$°K, $\Delta\nu = 1$ GHz. (2) $\lambda = 1$ m, $T_{max} = 10^{12}$ K, $T = 100$°K, $\Delta\nu =$ 100 MHz. (3) $\lambda = 1$ cm, $T_{max} = T$, $\Delta\nu = 1$ GHz. Solid horizontal lines indicate limits of angular resolution for various conditions. Solid inclined lines indicate limits of sensitivity corresponding to cases (1), (2) and (3).

elements. It is advisable to place the controlling element at the center of curvature. Since it is desirable that the radiation from the source under study be recieved simultaneously throughout the whole wavelength band, the schemes using a correcting mirror seem to be preferable (Head, 1957; Holt and Boushe, 1964).

Table 1 summarizes the fractional dimensions of a correcting mirror as a function of the fractional focal distance of the primary mirror (d, h, f are the diameter, depth, and focal distance of the correcting mirror, respectively; F is the focal distance of the primary mirror and is equal to about half its radius of curvature; and D is the diameter).

Asymptotic expressions for d/D, h/D and f/D are based on the results of Head (1957) (at $D/F \ll 1$):

$$d/D = \frac{1}{128}(D/F)^2 \tag{8}$$

$$h/D = \frac{3}{128}(D/F) \tag{9}$$

$$f/D = \frac{5}{3072}(D/F)^3. \tag{10}$$

Note that the ratio of the primary mirror diameter to the wavelength for such large antenna structures is of the same order of magnitude as for modern optical telescopes. Therefore, the design theory for optical telescopes with spherical reflectors can be used (e.g. Mihelson, 1976). To a first approximation, only third-order aberrations will be considered. Then the image dimensions at the focus of a sufficiently distant point-source will be determined by the diameter of the spherical aberration circle:

$$2\rho_{\text{I}} = \frac{1}{128}(D^3/F^2) \tag{11}$$

which coincides with the diameter of the correcting reflector (eqn 8). For a considerable angular deflection from the axis, image dimensions are determined by the coma dimensions; i.e. with the deflection at an angle θ from the axis, the size of the spot due to coma will be,

$$3\rho_{\text{II}} = \frac{1}{32}\frac{D^2}{F}\theta \tag{12}$$

Table 1. Dimensions of corrector mirror

F/D	0.4	1	2	4
d/D	0.05	0.008	0.002	0.0005
h/D	0.06	0.02	0.01	0.006
f/D	0.03	0.002	0.0002	0.00003
$\theta^0_{1/2}$	37	11	6	3

in a direction perpendicular to the primary reflector mirror axis, and,

$$2\rho_{II} = \frac{1}{16}\frac{D^2}{F}\theta \tag{13}$$

in the direction perpendicular to it in the image plane.

The image dimensions also depend on astigmatism, which stretches the point image in a direction perpendicular to the primary mirror axis up to the value:

$$2a = \frac{3}{8}D\theta^2. \tag{14}$$

Off-axis aberrations limit the field of view of such a telescope. Let us consider this problem with regard to the SRT.

Three illumination schemes can be used for spherical reflectors, providing multibeam reception:

(1) The primary mirror is partially illuminated by identical correcting mirrors at different angles θ to the axis, as in the Arecibo radiotelescope (Lowe, 1976). The efficiency of the primary mirror does not vary within the angle of view.

(2) The primary mirror is illuminated by identical correcting mirrors at different θ. The efficiency of the primary mirror depends on this angle (for each feed).

(3) The primary mirror is fully illuminated by several correcting mirrors, each having a shape optimal for a specific range of angles relative to the mirror axis.

The second scheme is more effective than the first since, at least near the primary mirror axis, the whole aperture of the reflector is used. Now we estimate how the antenna effective area is reduced when a correcting mirror moves along the circumference of $R_a/2$ radius through an angle θ relative to the center of the primary reflector (if perfectly uniform illumination with edge cutoff is assumed for co-axial location):

$$\eta = \frac{A_{eff}}{A_{eff_{max}}} = \left\{1 - \frac{2}{\pi}\sqrt{\left[1-\left(\frac{tg(\theta/2)}{tg\theta_{max}}\right)^2\right]}\frac{tg(\theta/2)}{tg\theta_{max}} - \frac{2}{\pi}\arcsin\frac{tg(\theta/2)}{tg\theta_{max}}\right\}\cos^2\frac{\theta}{2}. \tag{15}$$

Here $2\theta_{max}$ is the maximum aperture angle; $\sin\theta_{max} = r_a/R_a$, where r_a and R_a are the radii of the aperture and curvature of the primary mirror, respectively. The values of $\theta_{1/2}$, for which the effective area halves, are given in Table 1.

The third scheme makes it possible to use a large effective area when observing away from the primary mirror axis, but the dimensions of correcting mirrors become larger.

In the geometric optics approximation (for a feed lying on a radius of curvature at $R_a/2$, with a central angle θ relative to the primary mirror axis, and receiving radiation along this radius), it can be shown that the maximum dimensions of an image do not exceed the spherical aberration of some effective equivalent mirror. For this case, one must insert in eqn (11) the value $D_* = D \sin(\theta + \theta_{max})/\sin \theta_{max}$. Then, an upper limit for the fractional dimension of a corrector can be calculated:

$$\frac{d^*}{D} \leq \frac{1}{128} \frac{D^2}{F^2} \frac{\sin^3(\theta + \theta_{max})}{\sin^3 \theta_{max}}. \tag{16}$$

Table 2 gives this value for different θ and different F/D.

Table 2. Upper limit of corrector dimension, d^*

θ^0 \ F/D	0.4	1	2	4
20	0.12	0.09	0.1	0.13
10	0.08	0.04	0.03	0.03
5	0.07	0.02	0.01	0.007

A comparison of the second and third schemes shows that preference should be given to the second scheme, which is simpler. According to the data in Tables 1 and 2, the correcting mirror with $d/D = 0.8\%$, $F/D = 1$ and $\pm 10°$ field of view, at the periphery of which the effective area is reduced by a factor of 2, seems to be the best compromise. This version is a tradeoff between the requirements for a greater field of view and for smaller dimensions of the corrector.

A field of view of $\pm 10°$ corresponds to 1% of the area of the celestial sphere. Hence, full coverage of the sky will require only about 100 changes in primary reflector orientation.

The reflector can be extended at will if flat reflecting elements are used. Flat, fully symmetric hexagons with $b = 4$ m diagonal may be used as modules because they are easy to transport. Here, a spherical surface is approximated by a multifaceted surface consisting of flat identical elements; thus a regular error is introduced which distorts the wave front of an incoming signal and limits the value of the minimum useful wavelength. It can be shown (Braude *et al.*, 1960) that, with a 10% admissible loss in the effective area of the reflector, the minimum operating wavelength is,

$$\lambda_{min} = \frac{0.7 r_a^2}{R_a} \{1 + \sqrt{[1 - (r_a^2/R_a^2)]}\}. \tag{17}$$

If it is assumed that the shape of the primary mirror surface can be changed by adjusting the relative position of its elements, then there are three procedures for antenna extension throughout which the SRT can operate at all intermediate stages.

(1) Extension while keeping the aperture angle, θ_{max}, constant. In this case, the radius, R_a, grows proportionally with the aperture D and the minimum operating wavelength varies as $1/D$. This is considered a serious disadvantage of this system since observations at extremely short waves cannot be carried out

during the initial stages. The diameter of a correcting mirror, according to eqn (8), grows proportionally with that of the primary mirror; i.e. correcting mirrors should be changed (or extended) during the SRT main reflector extension.

(2) Extension while keeping the diameter of the correcting mirror constant. According to eqn (8), the sphere radius obeys the law $R_a \sim D^{3/2}$. Although the diameter of the correcting mirror is kept constant, its depth (eqn 9) grows proportionally with $D^{1/2}$, i.e. correcting mirrors (or their shape) should be changed during SRT extension. The minimum operating wavelength reduces following the law:

$$\lambda_{min} \sim D^{-3/2}.$$

(3) Extension while keeping the mirror radius of curvature, R_a, constant. Here the shape of the primary mirror surface should not be changed during extension; this facilitates the operation of the adjustment system (the range of required movements is smaller). The minimum operating wavelength remains almost the same ($\lambda = 11.0$ to 11.2 mm for $D = 0$ to 1 km, $R_a = 2$ km; $\lambda = 1.10$ to 1.12 mm for $D = 0$ to 10 km, $R_a = 20$ km). In this case, however, the diameter of the correcting mirror is variable and is determined by the sum of a diffraction disk dimension and the axial corrector diameter (eqn 8):

$$d_1 = \frac{\lambda R_a}{D} + \frac{1}{32}\frac{D^3}{R_a^2}. \tag{18}$$

Although during the initial stages of reflector extension there are restraints on the use of longer wavelengths, the last antenna scheme seems the most expedient.

With eqn (17), a mirror radius of curvature can be determined from the dimension of the flat elements and minimum wavelength. If $a = 4$ m and $\lambda_{min} = 1$ mm, then $R_a = 22$ km, which corresponds to $F/D \sim 1$ if $D = 10$ km.

The minimum operating wavelength is limited by the random errors of the reflector surface. According to Ruze (1968), the coefficient of efficiency of the surface, due to these errors, is,

$$\eta_a = e^{-(4\pi\sigma/\lambda)^2}. \tag{19}$$

Here σ is the rms error of the surface. If the loss $(1 - \eta)$ is 10% and if $\lambda = 1$ mm is assumed, the tolerable rms error σ is 0.03 mm; such should be the accuracy in manufacturing and adjusting the flat elements.

Now we turn to the requirements for accuracy of SRT orientation and stabilization. From $\psi = \lambda/D$, it follows that the antenna beamwidth becomes narrower as the antenna is extended. For a minimum wavelength of 1 mm and reflector diameters of 1 and 10 km, $\psi = 0.2$ and 0.02 sec of arc, respectively. To make measurements with such antenna beams, the positions of the feeds and their adjustment should be made also to the accuracy of the above value of σ. Note again that the accuracy of primary mirror orientation can be much lower, i.e., on the order of 1°.

4. Space radiotelescope design

The main element of an SRT is the antenna. The antenna structure should be able to preserve its integrity and stability under the influence of gravitational forces and light pressure. The required pointing accuracy also imposes restrictions on resonance and stiffness characteristics of the mirror. These restrictions and the service conditions determine the permissible minimum weight of the structure.

Two possible types of structures are considered: a solid, load-bearing shell, and a rod, load-bearing framework, on which a reflecting surface is mounted (Fager and Garriott, 1969; Crawford *et al.*, 1974; Hagler, 1976). Of interest also is a structure formed from a solid surface strengthened with stiffening ribs; but at present, suitable design solutions for SRT's are not available.

As can be shown, the maximum total normal load due to light pressure, $q_{N_{max}}$, may be determined from the relation:

$$q_{N_{max}} = 0.23\beta^2 q_0 \tag{20}$$

where β is half the central angle of the segment (in radians) and q_0 is the light pressure.

Because of the need for overall stability of the mirror system, $q_{N_{max}}$ should be less than the critical load, P_c, acting on the shell, which is (see Volmir, 1967):

$$P_c = 0.15E_n \frac{\delta_n^2}{R_a^2} \tag{21}$$

where E_n and δ_n are the reduced elastic modulus and shell thickness, respectively.

From the relation (21), the shell thickness can be determined and, consequently, the minimum reflector mass, M, which for small values of β, will be

$$M = 3.9r_a^3\rho\sqrt{(q_0/E)} \tag{22}$$

where ρ is the shell material density, E is the material elastic modulus, $r_a = R_a \sin\beta$ is the reflector radius, and R_a is its radius of curvature. For a double-lattice rod shell, whose basis element is a tetrahedron, the above characteristics can be written as (see Saveljev, 1966):

$$\delta_n = \sqrt{3}\,h_0; \qquad E_n = \frac{4EF_0}{3^{3/2}tg\alpha_0 h_0^2}; \qquad \rho_n = \frac{2\sqrt{3}F_0\rho}{tg\alpha_0 h_0^2} \tag{23}$$

where h_0 is the basic tetrahedron height, α_0 is the angle between the tetrahedron height and its rib, and F_0 is the rod cross-sectional area. In this case, the minimum reflector mass is,

$$M = 12.5\frac{r_a^4}{h_0}\frac{\rho_0}{E}q_0 + \pi r_a^2\rho_r \tag{24}$$

where ρ_r is the unit mass of the reflecting surface area.

In a rod structure, the minimum mass can also be determined from the condition for loss of stability by some of the rods. The maximum stress in a mesh-structure rod will be,

$$N = \frac{q_N l R_a}{4\sqrt{3}} \tag{25}$$

where l is the mesh rod length.

As is well known, the value of the critical stress for a thin-walled tubular rod is,

$$N_c = \frac{\pi^2 E F_0 d_0^2}{8l^2} \tag{26}$$

where d_0 is the rod diameter.

From these expressions, the minimum rod cross section and the mass of the load-carrying structure can be determined as,

$$M = 14\frac{\beta r_a^3 \rho q_0}{K_y^2 E} \tag{27}$$

where $K_y = 4d_0/l$ is the dimension ratio of the folded and deployed structure.

During reflector pointing and stabilization, the proper period of the control system, T_c, will be determined by the lower of the following two values:

$$T_c = \frac{2}{\pi} T^* \sqrt{(\Delta\theta/3)} \tag{28}$$

$$T_c = \sqrt{\left(\frac{r_a \delta_n \rho_n \Delta\theta}{q_0 tg\beta/2}\right)} \tag{29}$$

where T^* is the orbital period (of rotation) and $\Delta\theta$ is the precision of the reflector pointing. Expression (29) dominates for high orbits.

To avoid serious difficulties in designing the control systems, the period T_1 of the reflector's natural oblique-symmetric oscillation should be an order of magnitude lower than the period T_c.

The greatest period of a solid spherical segment's natural oblique-symmetric oscillation is,

$$T_1 = 1.7\frac{r_a^2}{h_0}\sqrt{(\rho/E)} \tag{30}$$

an approximate expression obtained by the energy method.

From $T_1 \leq 0.1T_c$, we obtain,

$$r_a/h_0 = (2\Delta\theta E/3\beta q_0)^{1/3}. \tag{31}$$

The minimum mass of the solid structure is

$$M = \pi r_a^3 \rho (3\beta q_0 / 2\Delta\theta E)^{1/3}. \tag{32}$$

The period of rod structure oscillation is determined from

$$T_2 = 0.52 r_a^2 \sqrt{(\delta_n \rho_n / D_n)} \tag{33}$$

where D_n is the normalized cylindrical stiffness of the load-carrying structure.

If it is assumed that $T_2 \leq 0.1 T_c$, we obtain the minimum mass as:

$$M = \pi r_a^2 \rho_r \frac{3 \times 10^{-17} r_0^3 \Delta\theta h_0^2 E}{3 \times 10^{-17} r_0^3 \Delta\theta h_0^2 E - r_a^4 \rho}, \tag{34}$$

or else,

$$M = 375 q_0 \frac{\beta \rho r_a^5}{\Delta\theta E h_0^2} + \pi r_a^2 \rho_r \tag{35}$$

for values T_c from eqns (28) and (29), respectively, and r_0 is the orbit radius.

According to the constraints induced by putting the spaceship in orbit, the maximum value of h_0 for rod structures is about 10 m. The value K_y for folded rod structures, available now, does not exceed 0.2. The mass/diameter dependence in such structures is given in Fig. 6 for the following values: $E = 1.5 \times 10^{11}$ Pa, $\rho_0 = 1.4 \times 10^3$ kg/m^3, $\rho_r = 0.1$ kg/m^2, $\beta = 15°$, and $\Delta\theta = 1°$ (Fager 1976). These curves show that, beginning from $r_a = 500$ m, the solid structural mass exceeds the mass of the rod structure by more than two orders of magnitude.

Rod structures providing a three-dimensional framework on which the reflecting surface is mounted are more effective than continuous load-bearing shells. Depending on the minimum wavelength, the reflecting surface may be made from metal mesh or from precision-fabricated panels, permitting adjustment of their position relative to the framework.

The use of collapsible structures will make it possible to minimize assembly

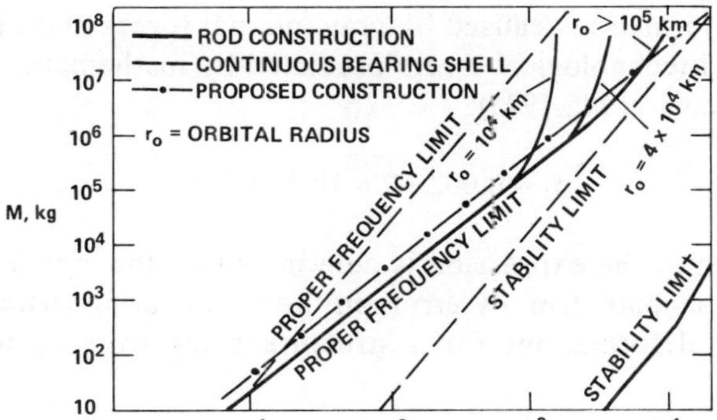

Fig. 6. Antenna mass, M, versus antenna radius, r_a.

work in orbit. However, the fabrication and single-shot launching of a collapsible structure with a 1-km diameter will hardly be possible in the near future. On the basis of projected transport facilities, the dimension of an antenna module injected into orbit will be 200 m. The folded modules are injected into orbit, then automatically deployed and docked together. In this situation, it is possible to increase the dimensions of the operating antenna reflector. The module framework consists of a primary structure and a reflector working surface on which the reflecting surface is mounted.

The above ratios show that minimum mass and maximum stiffness are achieved with the greatest height of the structure. Therefore, the load-bearing framework is made as a folded two-mesh shell with a maximum rod length, determined by transport facilities available. In this case, the structure height will be 10 m, which allows the antenna reflector to be increased to about 20 km (see Fig. 6). A still larger antenna will require much greater mass.

The surface of the load-bearing framework may consist of 15-m triangular cells. The working surface installed inside each cell is made as a single-mesh and rod structure with triangular sub-cells about 2 m on edge (Fig. 7). The diameter of the working surface rods is 1/10 the load-bearing framework rods because of the need for installing the working surface between the load-bearing frame assemblies.

The load-bearing frame is made from thin-walled tubes of 75-mm cross section and a wall thickness of 0.5 mm. The total number of rods in a load-bearing framework block is 1300, and the total mass of the frame is 4 tons.

If adjustable precision panels are mounted on the working surface, the reflecting surface mass will be approximately equal to that of the framework.

Depending on the dimensions and on the mode of operation, the interconnection between these modules can be either rigid or not, allowing for adjustment of their mutual positions. In such a configuration, these modules will be suitable for reflecting antennas in sizes up to about 20 km diameter and operating down to very short wavelengths.

During antenna operation, the maximum distortions of the reflecting surface should not exceed 1/16 of the minimum operating wavelength. Surface distortions will result from technological errors in fabrication, temperature deformations, and deformations caused by gravitational forces and light pressure.

The expected technological errors predicted by mathematic modeling can be found (Gvamichava *et al.*, 1975):

$$\delta_1 = 0.002 + 7\times 10^{-4}(2r_a^2/l)^{1/2}. \tag{36}$$

The first term of the expression is determined by the errors of the reflecting surface and the second term, by errors in the load-bearing structure.

Temperature deformations for a structure made from up-to-date materials will not exceed,

$$\delta_2 = 2\times 10^{-5} r_a. \tag{37}$$

Deformations caused by gravitational forces can be determined from an

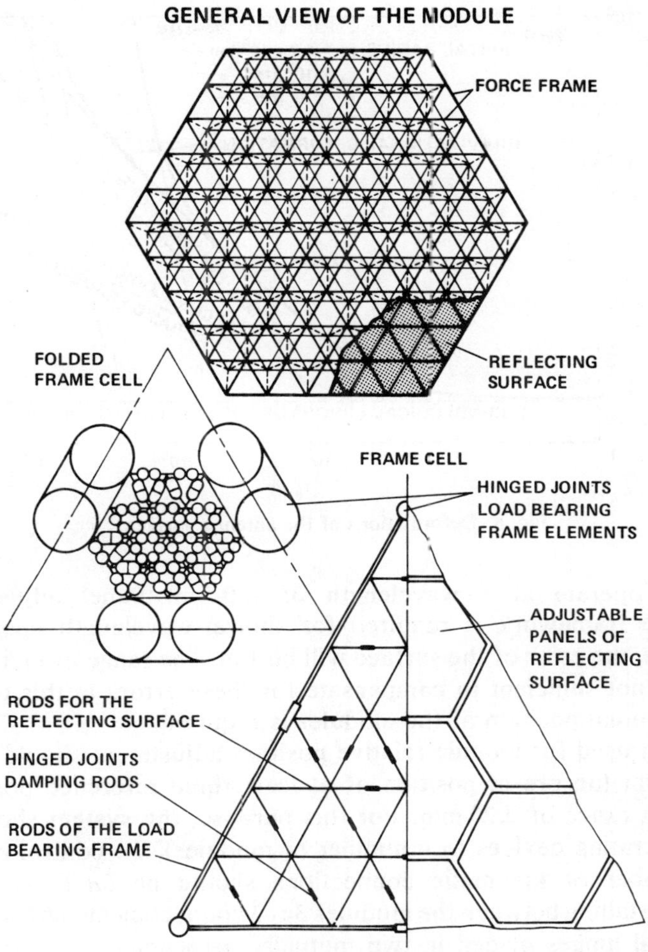

Fig. 7. Structure of antenna frame.

approximate expression obtained by the energy method:

$$\delta_3 = 0.8\beta\mu_0 \frac{\rho r_a^5}{E r_0^3 h_0^2}(1+K) \tag{38}$$

where μ_0 is the gravitational constant, K is the ratio between the reflecting surface weight and the weight of the framework.

Deformations caused by light pressure will be:

$$\delta_4 = 1.5 \times 10^{-2} \frac{\beta q_0 tg\alpha_0 r_a^4}{EF_0 h_0}. \tag{39}$$

These dependences and the total distortion of the surface are shown in Fig. 8. The curves show that, within one module, the error is about 8 mm. In a 1-km antenna, the error will be about 25 mm; i.e. without surface adjustments, the

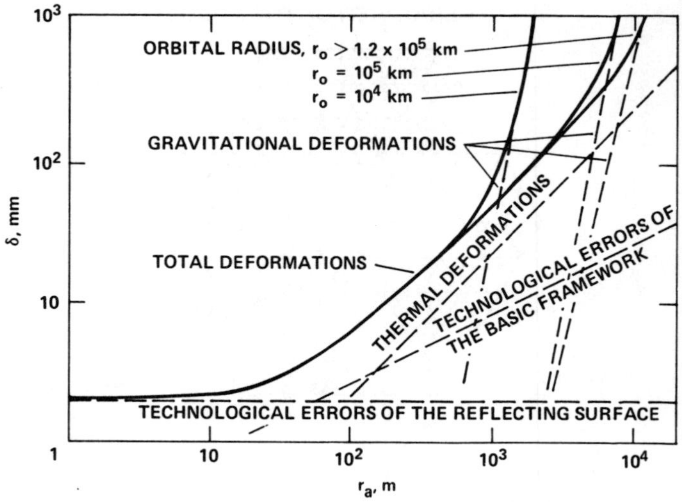

Fig. 8. Deformations of the antenna surface.

antenna can operate at a wavelength of ≥ 0.5 m. Panel adjustment with respect to the framework is required for shorter-wavelength operation. In a 10-km antenna, the error of the surface will be 1 m. The range of individual panel adjustment is not sufficient to compensate for these errors in this case; adjustment of the mutual position of the modules is required.

The system used for module relative position adjustment should provide for movement from the preset position of at least three reference points in each module over a range of ± 20 mm. For this purpose, the system should have at least $3n-3$ operating devices (n = number of modules). To control each device, the total number of kinematic connections should be $6n-6$, which can be achieved by installing between the modules $3n-3$ connecting members each having four cylindrical hinges placed in two mutually perpendicular planes, and by a device that can change the connecting member length. Such elements should be mounted between each pair of modules, but only two connections between edge modules.

With an increase in the antenna aperture diameter beyond 20 km, gravitational deformation should increase sharply. Also, with such dimensions, the natural oscillation frequencies are lower than tolerable.

An increase in antenna dimensions beyond 20 km will require a corresponding increase in the height, h_0, of the load-carrying structure; i.e. a three-dimensional extension of the structure will become necessary.

Consider now a few problems in the area of SRT control. The control system should provide for simultaneous operation of all members of the SRT. This includes orientation of the focal axis of the primary reflector, its rotation (Preston *et al.*, 1976), the fixed position of the receiving and controlling elements in the coordinate system of the primary reflector, as well as orbit correction of the SRT as a whole.

For a spherical primary reflector, the accuracy of its focal axis orientation should be at least 1°. The destabilizing moments of light pressure and gravitational forces will affect the accuracy of the mirror. The maximum gravitational moment is,

$$M_g = \frac{3\pi}{8}\frac{r_a^{\,4}\mu_0\delta_n\rho_n}{r_0^{\,3}}\frac{\sin 2\nu}{\cos^2\beta/2} \tag{40}$$

where, v is the angle between the focal axis and the directions to the center of the Earth.

The maximum moment due to light pressure, obtained by integration over the antenna surface, is,

$$M_r = \frac{\pi}{4}r_a^{\,3}q_0 tg\beta/2\phi(\chi) \tag{41}$$

where χ is the angle between the focal axis and the solar direction and

$$\phi(\chi) = \frac{1}{4}\sin 2\chi \operatorname{sign}(\cos\chi) + \frac{1}{3}\left(1 - \frac{1}{3}\sin^2\beta\right)\sin\chi.$$

The case of a continuous shell with diffuse light scattering is considered here.

The 1-km reflector orientation and stabilization can be achieved with several jet engines having a thrust of about 1 kg, mounted on the antenna periphery. Since the controlling angular accelerations are 10^{-7} to 10^{-10} sec^{-2}, unilateral cycles of oscillations can occur in the orientation system. Fuel consumption required for light pressure compensation is,

$$G_r = \frac{2\pi r_a^{\,2}t}{Ig}\Phi(x)tg\frac{\beta}{2} \tag{42}$$

where I is the thruster specific impulse and t is operation time.

If ion thrusters with specific impulse on the order of 10^4 sec are used (Grishin *et al.*, 1975), daily consumption of propulsive mass to compensate for the pressure of sunlight on a 1-km antenna will be about 0.25 kg (for $F/D = 1$); for the 10-km antenna, it will be about 125 kg. The power supply for thrusters with a specific thrust of 600 wt/g should be about 15 kW for a 1-km antenna, and about 1500 kW for a 10-km antenna.

Fuel consumption for gravitational compensationn is obtained similarly;

$$G_g = \frac{3\pi r_a^{\,3}\mu_0\delta_n\rho_n}{8r_0^{\,3}I}\frac{\sin 2v}{\cos^2 v/2} \tag{43}$$

For a 1-km antenna (with $M = 200$ tons) in a stationary orbit, the daily consumption of propulsive mass will be about 6.4 kg and power needed for the ion engines will be about 200 kW. For a 10-km antenna, these parameters will increase to 6.4 tons and 200 MW, respectively.

Fuel consumption for antenna rotation through an angle α^* during time t is (see Raushenbah *et al.*, 1974),

$$G_n = \frac{tp}{Ig}\left[1 - \sqrt{\left(1 - \frac{\pi\alpha^*\delta_n\rho_n r_a^{\,3}}{pt^2}\right)}\right] \tag{44}$$

where p is the thrust of the control engine and g is the acceleration of gravity.

One to 5 kg of fuel are required for the $\alpha^* = 180°$ rotation of a 1-km antenna during 24 hr; 1.5 tons of fuel are required for the rotation of a 10-km antenna through the same angle during 5 days.

Let us calculate the fuel consumption for position stabilization of the receiving elements. During an SRT positioning while in geostationary orbit, differential acceleration occurs between the receiving element and the primary reflector:

$$\mathbf{a} = \frac{\mu_0}{r_0^5}[3\mathbf{r}_0(\mathbf{l}\mathbf{r}_0) - \mathbf{l}r_0^2] \tag{45}$$

where $\mathbf{l}$ is the vector connecting the mass centers of the primary reflector and the other system under analysis. For an SRT with a focal length of 400–4000 m, the acceleration of the receiving element will be 6×10^{-4} to 6×10^{-3} cm/sec^2. When ion thrusters with a specific impulse of 10^4 cm are used (Grishin *et al.*, 1975) to compensate for these disturbances, a propulsive mass 6×10^{-6} to 6×10^{-5} of the mass of the element should be consumed daily.

The thrusters of the receiving system should also provide high accuracy when tracking an astronomical source during the occurrence of variable disturbances to the position of the primary reflector. Under the condition of unilaterality, cyclic angular disturbances with accelerations on the order of 10^{-7} to 10^{-10} sec^{-2}, and with a precision of primary reflector pointing on the order of 1°, these disturbances will require a total increase in velocity of 0.06 and 0.6 m/sec/day for the 1-km and the 10-km reflectors, respectively. In this case, fuel consumption will be 2×10^{-7} to 2×10^{-6} of the initial system mass; i.e. considerably less than is required to insure the motion of the system along the given trajectory.

Let us also consider the possibility of an SRT transfer from a low Earth orbit into an interplanetary orbit as is required to provide an interferometer base line, B, after the SRT assembly (see Section 2). It follows from the above analysis that a 1-km SRT can be assembled at a height of at least 1000 km and a 10-km SRT, not lower than at a geostationary orbit.

Injection into the working orbit can be provided only by low power engines with a thrust limited by the stiffness of the antenna structure. Let us estimate the engine system characteristics necessary to raise the velocity of an SRT in the initial circular orbit up to parabolic velocity. Using the equation for acceleration of a body moving with a constant thrust (Beletsky and Egorov, 1964) along a spiral trajectory, the propulsive mass consumption and time to parabolic velocity can be calculated. The specific impulse of the thruster was 10^4 sec, the acceleration for a 1-km antenna was 10^{-4} g (a restriction caused by the engine's power supply); for a 10-km antenna, the acceleration is limited to 3×10^{-5} g (a restriction because of the low structural rigidity). Calculated values for the time to final velocity and the relative propulsive mass consumptions are given in Table 3.

Table 3. Basic engine system characteristics

Height of initial orbit (10^3 km)	System diameter (km)	Time of velocity increase (days)	Approximate propulsive mass consumption (%)
1	1	75	7
	10	250	7
36	1	25	2.5
(geostationary)	10	95	2.5

5. Injection into orbit and feasibility of an SRT

Technical and economic estimates were made to determine a reasonable way to assemble an SRT in orbit and to judge the feasibility of the project.

To define how a telescope's parameters and the space vehicles used for its injection affect major technical and economic factors in an SRT project, it was worthwhile to determine only approximate weights and costs without detailed development of a realistic version. The estimates obtained should be considered as only approximate, for which the possible versions are characterized according to their relationship rather than according to absolute factors.

As a result of our preliminary considerations, we have chosen the following scenario for assembling an SRT in Earth orbit:

- Inject individual blocks and elements of the telescope (in transportable packages) into low orbit by a reusable chemical rocket.
- Assemble telescope building blocks into "trains" in a low orbit by means of a special orbital "tug".
- Transport "trains" into working orbits by interorbital tug with a nuclear electric jet engine.
- Assemble telescope in final orbit.

Men should take part in the telescope assembly. Therefore, it is necessary to have already provided a 10–15-man crew and a work station in the final working orbit. (Another prospect is to use robots that can accomplish SRT assembly and operation with ground commands.) The station is a base for antenna assembly. Transportation and exchange of crews and life-support systems are provided by manned, interorbital, chemical, expendable or reusable spacecraft (depending on the working orbit height).

When the telescope is operational, consumable items (mainly fuel for orientation, pointing, and for assembly in orbit) are to be delivered by a tug with nuclear engines.

It should be mentioned that during telescope assembly in relatively low orbits (up to thousands of kilometers), it seems to be more practical to use direct delivery of individual blocks into the working orbit or interorbital vehicles working on chemical fuel. From this point of view, the quantitative estimates given below for the telescope with a 1-km antenna in an orbit of some thousands of kilometers height can turn out to be overstated; but this does not affect the major conclusions made from our study.

When each project was estimated, we chose (1) adequate parameters for a reusable, low-orbit vehicle and (2) the interorbital tug, (3) a project for payload-crew vehicles going from Earth to high-orbit stations and back to Earth, (4) the optimal height of circular working orbit for the telescope—all of which depend on the telescope aperture diameter. Criteria we considered were: total expenditures, S_Σ, for the assembly and operation of the telescope; among them, the cost of telescope development and of the necessary vehicles for its assembly and operation, including technological tests; the cost of the telescope and space facilities development, and the cost of their injection into orbit and of their assembly; the cost of SRT operation, including expenditures for the delivery of consumable items and the operation of the command-instrumentation complex for 10 years. The expenditures on development of a tug were not considered if the parameters of the low-orbit reusable carrier corresponded to a payload mass of less than 30 tons. The following restrictions also were adopted: the time for telescope assembly should not exceed 5 years; the launches of low-orbit tugs should not be more than once in 3 days; the weight of the interorbital vehicle without payload and fuel should not exceed that of the low-orbit carrier payload; and the exchange of crews at the high-orbit station should be performed at 6-month intervals.

The necessary change in antenna stiffness (and weight, consequently) characteristics with a change in the working orbit height was taken into account according to the desired limits on deformations of its reflector due to the Earth's gravitational field. The results of the calculations are given as plots in Figs. 9–12.

Figures 9(A) and 10(A) show the change in total weight G_Σ and cost S_Σ of the radiotelescope as a function of the working orbit height, H, for 1- and 10-km-diameter apertures, respectively.

The injection cost of 1 kg of 1-km-aperture antenna in geostationary orbit (without the cost of development, assembly, and operation) was chosen as a conditional unit of cost. This unit was chosen to facilitate comparison of our results with those of other investigations related to the problem of injecting various payloads into high orbits.

The weight G_Σ is given for the low circular orbit of 220 km, and it characterizes the delivery of freight (antenna blocks, transport spacecraft, fuel, etc.) by reusable low-orbit carriers, on the direct route "Earth to low orbit".

Figures 9(B) and 10(B) illustrate the distribution of total weight G_Σ over the telescope proper (antenna and auxiliary elements), its injection in orbit and assembly (mass of tug, fuel, and propulsive mass for a high-orbit station, transport vehicles, the facilities for their launch, etc.), and over the operation (fuel mass, transportation facilities, etc.).

The distribution of the total cost, S_Σ, over the development, assembly, and operation is shown in Fig. 9(C) and 10(C).

Figures 9(A) and 10(A) demonstrate that, for each dimension of the telescope, there is an optimal height for the working orbit for which the total expenditures on its development, assembly, and operation are minimum. The variation in the optimum value for G_Σ and S_Σ, as they depend on antenna aperture diameter D, is given in Fig. 11.

These estimates show that expenditures in the range of $(1.5 \text{ to } 12) \times 10^6$

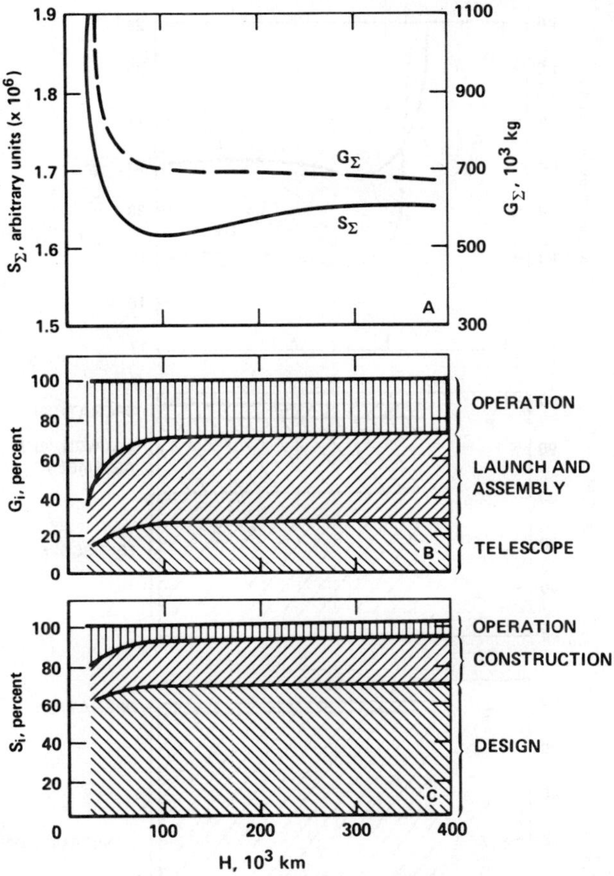

Fig. 9. Weight, *G*, and costs, *S*, for the 1-km radiotelescope assembly vs working orbit height; i = particular and Σ = total.

conditional units are required to develop and assemble large space radiotelescopes with antenna apertures from 1 to 10 km.

Assuming the value of a conditional unit as 500–750 (see, e.g. Koelle and Koelle, 1976), the expenditures in absolute units can be assessed in $\$10^9$ as a function of antenna aperture diameter.

Expenditures on the operation of such a telescope per decade can be assessed as 3 to 8% of their development and assembly cost.

The expenditure comparison between the development and assembly (without operation) of space and ground-based telescopes based on the present results and other data (Oliver and Billingham, 1973) (see Fig. 12) makes it possible to assume that the cost of assembling space telescopes with antenna aperture diameters of some kilometers will be lower than the construction costs of systems of ground-based radiotelescopes with the same effective area. A similar conclusion is given by Sheaffer (1976), who notes that, with an antenna of about 5 km, the cost of space-telescope assembly will be half the cost of constructing a ground-based system having the same equivalent diameter.

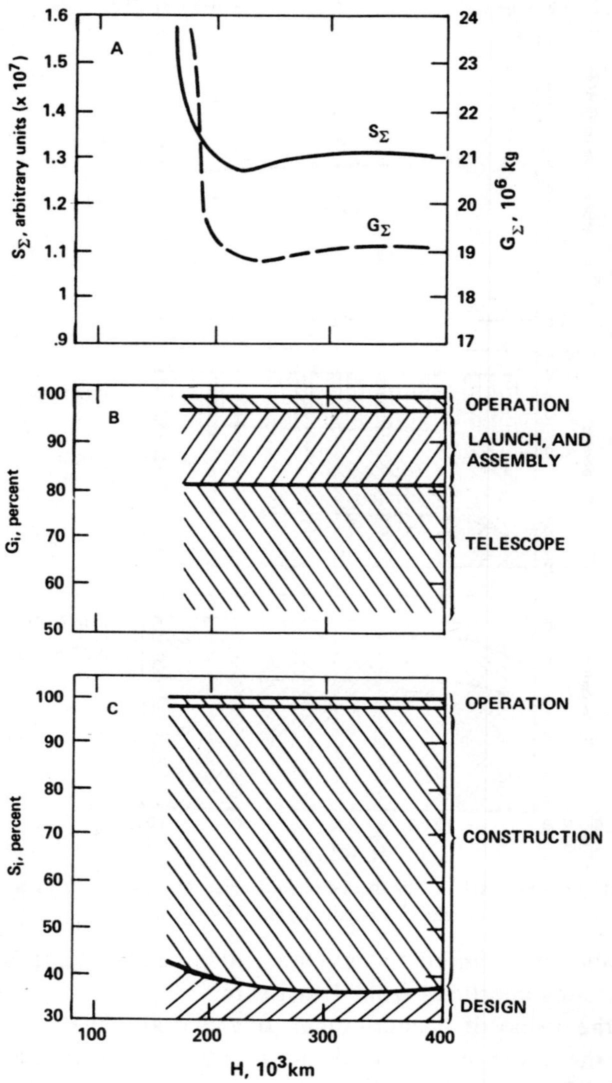

Fig. 10. Weight and costs for the 10-km radiotelescope vs working orbit height; symbols as in Fig. 9.

Conclusions

However difficult space activities are, the Cosmos provides many unique possibilities for astronomical studies that cannot be carried out on Earth. The calculations and estimates made show that space radiotelescopes will allow large-scale studies of the Universe up to distances on the order of its radius of curvature. Most urgent problems could be solved, including those of the evolution of life in the Universe, the formation and evolution of some cosmic objects (stars and planetary systems, galaxies, and quasars). All this is possible since, under conditions of zero-gravity, far from Earth, very large antennas can be

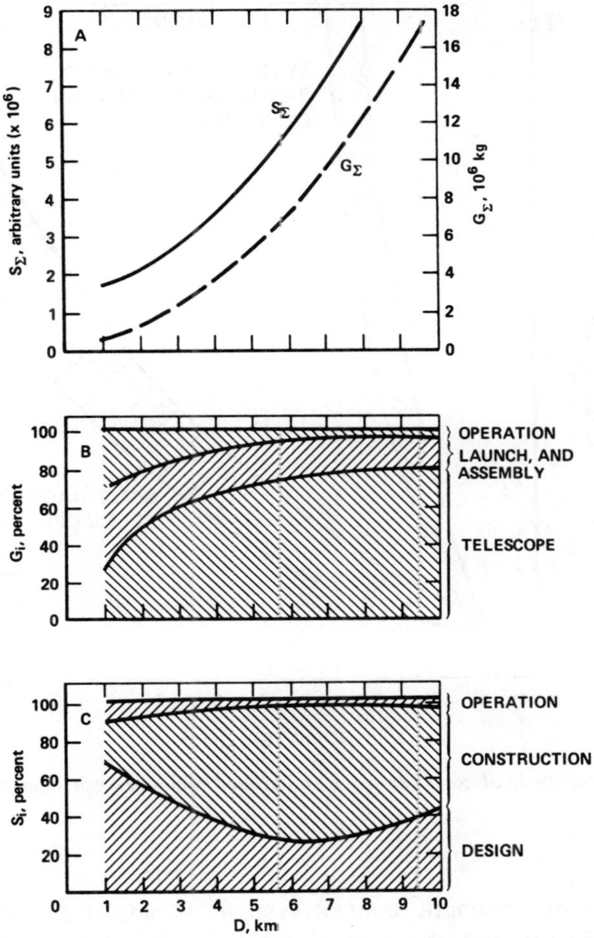

Fig. 11. Weight and costs for radiotelescope assembly vs antenna diameter for optimal orbit height; symbols as in Fig. 9.

built. Strong radio interference, typical for ground-based systems, and atmospheric effects can be eliminated, and radiotelescopes may be spaced by distances on the order of that of the solar system.

It should be mentioned that technological problems in connection with SRT development are similar to those of modern space technology; for both purposes, reusable Shuttle-type spaceships should be developed, as well as the principles for constructing very large space systems. Similar requirements arise in discussing projects of large space solar power stations and large research stations. In SRT development, as well as in these projects, a number of complicated technological problems should be solved:

- Delivery of large systems into space, their assembly, and tugging into circumsolar orbits.

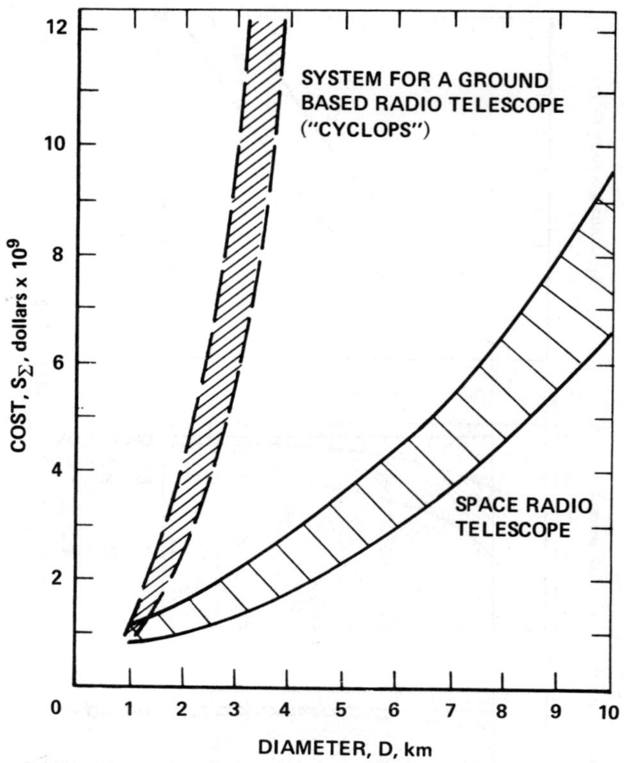

Fig. 12. Comparison of costs for space and ground radiotelescope construction.

• Development of automatic control systems to adjust the surface of the primary SRT reflector and the mutual position of receiving elements with accuracies required for millimeter-wave operation.

• High-accuracy measurements of SRT position; accurate determination of radio signal propagation time; transmission of broadband signals and their processing.

All these problems are being studied.

The estimates made show that it is basically possible to assemble space radiotelescope reflectors about 20 km in diameter in Earth orbits, using automatically deployed antenna modules about 200 m in size. To operate such antennas in the short-wave radio region (down to and including millimeter waves), automatic systems are required to adjust the primary reflector surfaces.

The specific mass of such antennas will be 0.2 to 0.3 kg · m^{-2}. The total cost for developing an SRT and putting it into orbit can be assessed as (\$1 to \$10) × 10^9 for reflectors 1 to 10 km in diameter.

The development and testing of a module for an extendable space radiotelescope, i.e., an SRT with a reflector about 200 m in diameter, can be regarded as a problem to be attacked now.

References

Anon. (1974) Programma CETI. *Astron. J.* (*USSR*) **51**(5), 1125.

Anon. (1977) *Sky and Telescope* **53**(6), 428.

Beletsky V. V. and Egorov V. A. (1964) *Komicheskie issledovanija* **2**, 392.

Braude B. V., Esepkina N. A., Kajdanovsky N. L. and Hajkin S. E. (1960) *Radiotechnika i Elektronika* **5**, 585.

Crawford R. F., Hedgepeth J. M. and Preiawerk P. R. (1974) *AIAA Paper* 1267.

Fager J. A. (1976) *AIAA Paper* 238.

Fager J. A. and Garriott R. (1969) *Trans. IEEE* **AP-17**(4), 452.

Grishin S. D., Leskov L. V. and Kozlov N. P. (1975) *Elektricheskie raketnie dvigateli.* Mashinostrojenie, Moskwa.

Gvamichava A. S. *et al.* (1975) *Report on All-Union Conference "Nadeznost v stroitelnoj mehanike".* Moskwa (thesis).

Hagler T. (1976) *Astronaut. Aeronaut.* **14**, 56.

Head A. K. (1957) *Nature* **179**, 692.

Holt F. and Boushe E. (1964) *Trans. IEEE* **AP-12**, 44.

Kardashev N. S., Parijsky Yu. N. and Sokolov A. G. (1971) *Uspehi Fizicheskih Nauk* **104**, 328.

Kardashev N. S., Parijsky Yu. N. and Umarbaeva N. D. (1973) *Izvestija SAO* **5**, 16.

Kellermann K. and Pauliny-Toth I. (1969) *Astroph. J.* **155**, L71.

Koelle D. E. and Koelle H. H. (1976) *Raumfahrtfohrschung* **20**, 57.

Lowe A. (1976) *Radio Science* **11**(8, 9), 671.

Mihelson N. N. (1976) *Opticheskie teleskopi.* Nauka, Moskwa.

Oliver B. M. and Billingham J. (1973) *Project Cyclops: a Design Study of a System for Detecting Extraterrestrial Intelligent Life.* NASA CR-114445 (Revised Edition).

Preston R., Hagar H. and Finley S. (1976) *BAAS* **8**(4), 497.

Raushenbah B. and Tokar E. (1974) Upravlenije orientatsiej Kosmicheskih apparatov. Nauka, Moskwa.

Ruze J. (1968) *Proc. IEEE* **54**(4), 633.

Sagan C. (Editor) (1973) *Communication with Extraterrestrial Intelligence,* p. 352. M.I.T. Press, Cambridge, Mass.

Saveljev V. A. (1966) In *Metallicheskie Konstruktsii.* Strojizdat, Moskwa.

Sheaffer R. (1976) *Spaceflight* **18**(10), 343.

Volmir A. S. (1967) *Ustojchivost deformiruemih sistem.* Nauka, Moskwa.

Acta Astronautica. Vol. 6, pp. 203–212. Pergamon Press 1979. Printed in Great Britain

On active and passive CETI from an earth satellite orbit

MIECZYSLAW SUBOTOWICZ†
Institute of Physics, M. Curie-Sklodowska University, Lublin, Poland

AND

JERZY USOWICZ†
Institute of Astronomy, M. Kopernik University, Torun, Poland

AND

ZBIGNIEW PAPROTNY
Polish Astronautical Society, Katowice, Poland

Abstract—Scientific and technical problems of the project (Subotowicz, 1976) were discussed in detail (e.g. single- or multiantenna system, frequency range, kind of orbit, corrections to antenna shape, role of thermal heating, gain of antenna, *possible perturbations of the orbit, undesirable motion of the antenna*, possibilities of long-term and exact pointing at a particular star, optimal dimensions of the antenna for different frequencies and distances in active and passive communication, modulation modes, etc.). Problems of the CETI antenna coupling with the solar power station (noise level, screening of the antenna from the power station, role of *antenna mechanical oscillations*, etc.) are discussed, as well as the shape and nature of the beacons and that of the S/N ratio; the laws of pulse distortion in the interstellar medium, and the role of Doppler shift of the signal.

1. Introduction

In 1976, Subotowicz proposed the concept of a satellite CETI transmitter and receiver in the form of a huge radiotelescope, orbiting synchronously and coupled with a satellite solar power station delivering 10–20 GW electrical power. It was accepted that the optimal range of the transmitted or received frequencies for CETI need not necessarily be equal and in the range between 1 and 2 GHz, chosen because of Earth's atmospheric absorption.

All transportation facilities together with the Space Shuttle and technology necessary to place the huge CETI radiotelescope in orbit are the same as that required to build a space solar power station (SSPS).

Few radiotelescopes are now used for CETI because of lack of time. Thus now is the proper time to build special radiotelescopes for CETI only. For a

†Polish Astronautical Society, Warszawa, Poland.

CETI search conducted from the Earth's surface, the microwave region of the electromagnetic spectrum is the most promising (Oliver and Billingham, 1973). Using orbital CETI facilities, one can work in the UV, IR, visible and millimeter wave regions. It is highly probable, however, that one type of CETI system will not be able to combine passive and active communications, both on Earth and in orbit. The program presented by Subotowicz (1978) can be considered as complementary to the Cyclops program.

An orbital CETI system as proposed by Subotowicz has two main advantages: (a) availability of the huge energy resources of the SSPS and the possibility of exploiting all transportation and installation methods and tools used for SSPS; in that case, an active CETI would be independent of terrestrial energy sources. (b) A message could be sent in more directions than in the case of transmission from the Earth's surface, assuming only one hemisphere is accessible to one radiotelescope.

The signals produced by terrestrial transmitters can interfere with orbital CETI reception since the sensitivity of the detectors should be about (10^{27}–10^{30})$WHz^{-1}\,m^{-2}$.

The cost of a radiotelescope built on the Earth's surface is determined by its weight and dimensions, and is strongly related to the admissible deformation caused by gravitation, winds, and temperature. In space, these do not exist, except for the thermal one. Radiation pressure and tidal forces may play a negligible role. Probably, in 10–20 years, the costs of constructing radiotelescopes on the Earth's surface and in space will be comparable. Space will have the tasks that cannot be realized on Earth because of the opacity of its atmosphere.

2. Search strategy

2.1. *General principle*

When realizing an active CETI, we can apply Dixon and Cole's strategy (1973, 1977). It takes into account the following dimensions unknown to both civilizations: distance; direction; polarization; signaling rate, bandwidth, and modulation, all interrelated; and frequency.

When planning CETI, we should minimize (Dixon, 1973) the extent of these dimensions by applying ideas of *anticryptography*... "beacons should be designed and operated in such a way as to maximize their probability of discovery, both by intentional searches and by accidental observation", and *mediocrity*... "we are just average among the inhabitants of the universe". One of us (M.S.) proposes a third idea, *partnership*: "Every civilization realizes active and passive CETI if only it can do it".

2.2. *Antenna system and monitoring*

It is important to continuously monitor the entire sky. Therefore, for a passive orbital CETI, the receiving system should consist of many antennas with large collecting areas. An ideal solution would be antennas built on the principle of the Luneberg lens, enabling simultaneous tracking of many stars. Probably it will not be possible to build such a single system with present-day technology. One antenna having a great collecting area for ETI search and also of high directivity and sensitivity, is seriously limited in time.

Let us consider, according to Gerritsen and McKenna (1975), a time interval of T years divided into shorter observational intervals t, where t is the time for monitoring one star, equal, for instance, to 500 s. Let civilization 1 transmit during m_1 randomly distributed time intervals t in the direction of civilization 2. During the rest of the $(n - m_1)$ time intervals, the signals are transmitted in the direction of other stars, etc. During the $(n - m_2)$ time intervals, other stars are monitored.

Let $P_n^{m_1 m_2}$ be the probability that civilization 1 receives at least one signal from civilization 2. Suppose that we monitor 10^4 stars and each civilization uses just one antenna. Let two telescopes be devoted to CETI without break; then $m_1 = n/10^4$ and $m_2 = n/10^4$. For the first contact, $P_n^{m_1 m_2} \approx 1$, then $n \approx m_1 \cdot m_2$. One can easily show that, in this case, $T \approx 1600$ years, which means that first contact occurs after 1600 years. The situation is even worse when the radiotelescopes do not work continuously. Let civilization 1 transmit 10 days in 1 year and civilization 2 monitor continuously. Then $T \approx 5.8 \times 10^4$ years; i.e. first contact occurs after about 60,000 years—a time comparable to the supposed lifetime of the civilization!

The only reasonable solution is to build antenna systems that allow the simultaneous tracking of many stars. An orbital CETI system should consist of many antennas having large collecting areas and great sensitivities. Such a system will have reasonable time characteristics.

Our conclusion concerning the time and mode of transmission and reception is as follows: (1) If the probability of receiving signals is to be large, we should expect that an ETI will send its message continuously and omnidirectionally—and so will ETI think when looking for our signals. It means that we should send our beacons omnidirectionally and continuously and every scientific-technical civilization (STC) should do so. Why should we expect that somebody is sending beacon signals in such a way that other ETI will receive them with the largest possible probability, if we are not intending to do so? We and every other ETI are morally obligated to realize together *active* and *passive* CETI, either from orbit or from the surface of a planet. The active transmission of messages should be carried out with the largest possible power, transmitting the most understandable and readable information. (2) When a passive SETI is considered, we can accept that, at present, it is difficult to monitor many stars continuously. Therefore, taking into account the contemporary level of our STC, we should aim at the highest sensitivity and longevity of effort when monitoring the sky and scanning particular stars. Nevertheless, continuously monitoring the entire sky is the best method for SETI. This aspect of the strategy of CETI and SETI, i.e. the problem of the coincidence of the arriving signal and listening receiver, may be much more important than the uncertainty of frequency range or other technological details.

For CETI transmissions, the best solution would be to generate broadband and omnidirectional signals of very high power (about 10^{15} W) in the gigahertz range. Narrow-band pulses could also be sent to globula clusters or to galaxies, but this kind of strategy, because of the great distances, would mean a monolog by one civilization, not a dialog or conversation between many civilizations.

2.3. *Frequency range*

Serious arguments support the use of frequencies in the "waterhole" range between the hydrogen line (1.420 GHz) and hydroxyl line (1.662 GHz), when looking for ETI signals from the Earth's surface (through the atmosphere), and when working in the microwave region. For an orbital CETI, the optimal range of frequencies may extend from 1 to about 3000 GHz, i.e. from the microwave region to the infrared region (Subotowicz, 1976). In the first phase of SETI, the strategy of the isotropic and continuously working transmitter and isotropic and continuously working receiver should be used in the range of the minimum cosmic background, 0.8 to 50 GHz.

The width of the waterhole corresponds to 242 MHz. It is not possible to transmit CETI signals in such a broad band in the microwave region. In that case, the most suitable choice would be the hydrogen line, 1.42 GHz, for the following reasons: (1) it is the most common element in the Universe, (2) it is the strongest of all known lines, (3) the receivers for that frequency are built early in the development of radioastronomy, and (4) every civilization investigating the sky in the microwave region should know the hydrogen line.

2.4. *Modulation*

The only suitable modulation for signals intended for interstellar communication is two-state circular polarization modulation: left and right circularly polarized waves. This type of modulation is very convenient when using a binary code.

2.5. *Single antenna*: *wavelength* (λ), *diameter* (D), *and gain* (G)

Let us now calculate how the gain of an antenna depends on its diameter and operating wavelength. The relationship between λ, D and G is,

$$G = \eta G_{max} = \eta \frac{4\pi A}{\lambda^2} = \eta\pi^2 D^2/\lambda^2 = \eta(\pi D/\lambda)^2$$

where the area of the antenna is $A = \pi D^2/4$; G_{max} is the maximum value of G. Let the antenna efficiency coefficient, η, be 50%. (Antenna gain is often given in decibels (dB), G_{dB}.) From the definition, we have,

$$G_{dB} = 10 \log G.$$

To illustrate how G and G_{max} depend on λ and D, we have calculated G_{max} (Table 1) for several wavelengths and several antenna diameters. From Table 1, G and G_{dB} can be easily calculated for different λ and D.

2.6. *Earth-based antenna costs*

The cost of a parabolic radiotelescope depends on its diameter and the precision of its collecting surface. When telescopes are built on Earth, their cost (S) depends on the paraboloid diameter (D) and its area (A) in the following way:

$$S \sim D^{2.8} \sim A^{1.4}.$$

One square meter of collecting surface (s) costs $s = S/A \sim A^{0.4}$. The optimal size of an antenna can be calculated from a parametric, multi-dimensional analysis of the problem, and this is open for further development.

Table 1. Values of G_{max} for different wavelengths (λ) and different antenna diameters (D)

$D(m)$	$\lambda = 1$ cm	$\lambda = 10$ cm	$\lambda = 100$ cm
1000	9.86×10^{10} (110 *dB*)	9.86×10^{8} (90 *dB*)	9.86×10^{6} (70 *dB*)
1200	14.2×10^{10} (111.5 *dB*)	14.2×10^{8} (91.5 *dB*)	14.2×10^{6} (71.5 *dB*)
3000	88.8×10^{10} (119.5 *dB*)	88.8×10^{8} (99.5 *dB*)	88.8×10^{6} (79.5 *dB*)
5000	246.7×10^{10} (124 *dB*)	246.7×10^{8} (104 *dB*)	246.7×10^{6} (84 *dB*)

2.7. *Thermal deformations in space antenna*

Thermal deformations are the main source of performance degradation in large antennas orbiting the Earth because of the different incident angles of the solar radiation falling on the surface of the paraboloid; differential temperatures (ΔT) exist between points on the collecting surfaces. As a result, there exists a minimum wavelength (λ_{min}) for a particular ($D = $ const) radiotelescope because of this thermal deformation.

For paraboloidal radiotelescopes made of steel,

$$\lambda_{min} = 6(mm)\,\frac{D(m)}{100m}\cdot\frac{\Delta T(°K)}{°K}.$$

(The above relation is not true for a flat microwave reflector.) In Table 2, we present some calculated values of λ_{min} for different D and ΔT. As we see from the table, thermal deformations determine the optimum size of the antenna and the minimum useful wavelength.

Table 2. Minimum wavelengths λ_{min}(cm) for different radiotelescopes of diameter $D(m)$ and different temperature differences ΔT(°K) between points of collecting surface

ΔT(°K)	$D = 1000$ m	$D = 1200$ m	$D = 3000$ m	$D = 5000$ m
1	6	7.2	18	30
5	30	36	90	150
10	60	72	180	300

3. Doppler shift correction

It is generally accepted that ETI will transmit their signals in the vicinity of the 21-cm hydrogen line (rest frequency = 1.4204056 GHz). Because of the Doppler effect, the relative motions of the ETI transmitter and our civilization cause a frequency uncertainty of several MHz (our galactic rotational velocity is 250 ± 25 km/s). We cannot search this frequency range simultaneously. Therefore, Dixon (1973) and Dixon and Cole (1973, 1977) proposed to choose the Galactic center as the common reference point (Galactic Standard of Rest). In this reference frame, there exists a unique frequency for any given line of sight with respect to Doppler shift. One can assume that the transmitting civilization operates an omnidirectional beacon using this Doppler strategy, which means to transmit on a different frequency in each direction.

It seems that there is another possible approach to Doppler shift. We propose to use non-sine-like electromagnetic signals for interstellar communication. This kind of signal can be generated with a Walsh antenna. The time dependence of the current in the Walsh antenna is given by the Walsh functions. The Walsh system consists of complete, orthogonal, and orthonormal functions. Walsh antennas differ fundamentally from the Hertz dipole. The most important difference is the Doppler invariance of the Walsh antenna signals. Because of Doppler shift, the representatives of ETI's will use, for interstellar communication, electromagnetic signals invariant relative to the Doppler frequency shift. Walsh's waves can be generated using switching TRAPATT-diodes. Walsh functions have only two possible values, $+1$, or -1, and therefore are very convenient for binary systems.

Trigonometric functions are characterized by frequency; the Walsh functions through the number of zero passings (zps) and the time base, where the Walsh system is defined. As a result of the Doppler effect, the sine-like function is transformed into other sine-like functions of different frequency, depending on v/c. Meanwhile, the Walsh function transforms into the other only for $v/c \geq 3/5$. We should emphasize that Walsh functions are especially suitable for radio communication systems behind Earth's atmosphere. They permit better resolution, even exceeding the diffraction limit.

4. Compensation for dispersion effects

Signals sent by a particular civilization will be dispersed in space by free electrons. For the required time resolution $\Delta t(s)$, the maximum bandwidth (Δf_{max}) of the receiver for these signals is given by:

$$\Delta f_{max} = \Delta t \frac{df}{dt} \approx 1.2 \times 10^{-4} \Delta t f^2 (DM)^{-1}$$

where df/dt is the signal drift rate (in MHz/s), f is the monitoring frequency, and $DM = \int n_e dl$ (in pc/cm^3) is the dispersion measure. Let us suppose that $f =$ 1420 MHz, $DM = 50$ pc/cm^3:

For $\Delta t = 10^{-1}$ s: $\Delta f_{max} = 687$ MHz

For $\Delta t = 10^{-2}$ s: $\Delta f_{max} = 68.7$ MHz.

If the bandwidth of the high-frequency amplifiers is 200 MHz, only in the second case will dispersion compensation be necessary. The role of dispersion in the first case is negligible.

5. Pointing accuracy

For large antennas, positioning and tracking a particular point in the sky is more difficult than for small antennas. The tracking accuracy depends on the half-power beamwidth (HPBW) of the antenna pattern and on the rotation velocity of the sky ($V = 15 \cos \delta$, where δ is the declination of the star); $\text{HPBW} = 1.02\ \lambda/D$ radians. It follows from radioastronomical practice that an accuracy of 1/3 to 1/10 of the HPBW of the antenna is satisfactory, as well as a tracking accuracy of 1/10 to 1/50 HPBW. It means that the pointing accuracy for the antenna of diameter $D = 1.2$ km is 14.3–4.3 sec of arc, and the tracking accuracy, 4.3–0.86 sec of arc. It would be rather difficult for a huge orbiting radiotelescope to achieve this accuracy.

6. Length of signal path

The distance over which a signal may be satisfactorily communicated depends on several parameters, not only those of the message, but also on the probability of receiving errors, acceptable signal/noise ratio (SNR), etc. After many simplifications, distance L, where $\text{SNR} = 1$, can be written:

$$L = (P_t/Pr_{\min})^{1/2} \cdot \pi D_t D_r/(4\lambda).$$

For $D_t = D_r = 300\ \text{m} = 3 \times 10^4\ \text{cm}$, $\lambda = 21$ cm, $P_t = 0.25\ \text{MW} = 2.5 \times 10^5$ W, $P_{r_{\min}} = 3.2 \times 10^{-22}\ \text{W} \times \text{Hz}^{-1}$, the bandwidth $\Delta f_r \approx \Delta f_t$,

$$P_{r_{\min}} = 3.2 \times 10^{-22}\ \Delta f(\text{W})$$

$$L = 10^3/\sqrt{(\Delta f)}\ \ (1.y).$$

Indices r and t mean, respectively, receiver and transmitter; P is power, D is antenna diameter, and λ is wavelength.

7. Transmission duration in one direction

The amount of information (N bits) sent during t (seconds) in bandwidth Δf at $\text{SNR} = \alpha$ is given by:

$$N \approx (3/2)\Delta f \alpha t\ \text{(bits)}.$$

If the information contained in the beacon has N bits, then,

$$t = 2N/(3\Delta f \alpha).$$

For $\Delta f = 10$ Hz, $\alpha \equiv \text{SNR} = 1$, and $N = 10^3$ bits, $t \simeq 1$ min. Scanning time, T, depends on the width of the main lobe $2\theta_{1/2}(= \text{HPBW})$, distance, L, of the scanned star, and angular velocity, ω, of the scanning antenna.

$$T = t_2 - t_1 = f(L2\theta_{1/2}/\omega).$$

For an antenna of diameter $D = 300$ m and $\lambda = 21$ cm, $2\theta_{1/2} = 7 \times 10^{-4}$ rad.

8. Transportation problems, energy supply by solar satellite power station (SSPS)

As pointed out by Subotowicz (1978), transporting to satellite orbit, constructing, and assembling a huge radiotelescope, are based on the technology used to build the orbiting SSPS (Subotowicz, 1976; Gohrbandt and Schmidt, 1976). First-generation Space Shuttles will carry a useful payload of about 30 tons and transportation costs to low satellite orbits will be comparatively low, \$200–\$250/kg. Large second-generation Space Shuttles will carry useful payloads of up to 180 tons for about \$125/kg. SSPS will be built using Space Shuttles (Subotowicz, 1976; Gohrbandt and Schmidt, 1976). Exactly the same modular technology as used for the SSPS elements and their transport to the satellite orbit will also be used to build a CETI installation.

In synchronous orbit, space objects of large size and mass can be subjected to various perturbations because of the gravitational-field gradient, radiation pressure, interaction with the Moon and Sun, nonsphericity of the Earth, and inhomogeneity of the mass distribution in the Earth's crust. Transportation costs for the SSPS without the Space Shuttle are now \$1600/kw; when the Space Shuttle is used, the cost is \$500/kW. An SSPS of mass 30 ktons, delivering to the Earth 5 GW power during 30 years, would cost about $\$13 \times 10^9$. It would use a modular hexagonal panel (MHP) of photovoltaic cells. The transportation and installation costs in synchronous orbit would be 31% of the overall costs, the photovoltaic modules, 51%. We suppose here a conversion efficiency of about 18%. The SSPS, like the CETI installation, should be based on the modular concept to satisfy the following requirements: (a) cheap mass production of many modules, (b) low transportation costs, (c) easy assembly in orbit, (d) easy exchange of defective and/or degraded modules, and (e) possibility to easily enlarge the panel area to compensate for degradation losses during lifetime or to increase the power of SSPS or CETI installation parameters.

For the SSPS, several different arrangements of photovoltaic cells are proposed (Gohrbandt and Schmidt, 1976), based on an integral assembly of the array in orbit and on modular manufacturing on Earth and assembly of these completed modules in orbit (integrated spare panels, ISP; integrated rectangular panels, IRP; modular hexagonal panel, MHP; and modularized rectangular panel, MRP). One of the projected SSPS of 5-GW capacity based on the MRP concept consists of 55,000 modules and results in a weight of 3×10^7 kg (Gohrbandt and Schmidt, 1976). Its basic characteristics are given in Table 3. For the SSPS, a solar array area of about 70 km^2 would be necessary, assuming 10% overall efficiency.

Table 4 lists the weights and costs of the subsystems included in the overall weight and cost estimation of Table 3.

It is generally accepted that, after 30 years use, the silicon solar cell

Table 3. Principal characteristics of the 5-GW SSPS, based on the MRP concept (Gohrbandt and Schmidt, 1976)

Solar radiation intensity	1.4 kW/m^2
Solar cell efficiency	18%
Solar array concentration factor	2.0
Solar array output	8.0 GW
Overall transmission efficiency	62.5%
Power output on the ground	5.0 GW
Transmitting antenna area	1.0 km^2
Rectantenna area (collection and rectification system on the ground)	42.0 km^2
Overall cost estimation	\$13 × 10^9
Overall weight estimation	30.2 × 10^6 kg

Table 4. Distribution of weight and cost for 5-GW SSPS (MRP concept) (Gohrbandt and Schmidt, 1976)

Subsystem	Weight, %	Cost, %
Solar collector modules	77	51
Supporting structure	6	1
Transmitting subsystems	7	2
Receiving subsystem	—	4
Attitude control subsystem	8	3
Transport and installation	—	31
Maintenance (30 years)	—	6

conversion efficiency will degrade to about 1/3 its initial value. Therefore, it will be necessary to provide arrangements whereby solar array modules of differing performance can work well together.

For optimal delivered power/SSPS mass ratio, the voltage should be high: 60–80 keV. But modern microwave generators require a lower voltage of about 40 keV. A large number of cells will be connected in series and then in parallel to deliver the required voltage of about 40 keV and the required power of several GW. The power must be distributed to the microwave generators, by way of either single-bus or a multi-bus arrangement. In the single-bus system, all solar array modules feed the power into a main bus and the microwave generators are connected to the common DC cable via switching devices. This common cable, conducting 100-kA current, should be cooled with liquid helium down to 20°K, allowing a current density of about 100 A/mm^2 for copper, which means a cross section of about 1000 mm^2 for the conductors inside the cable. To compensate the strong magnetic fields produced by 50–100 kA, the cable should contain the forward line and return line. As the main cable length is 15 km long, the overall cable weight will be about 375 tons. The cooling machines require 45 MW and have a mass of about 250 tons. The switching gear required to start up the whole system consists of

10,000 switches for the 10,000 groups of 100 microwave generators each, and its overall weight is about 150 tons.

More reliable would be the multi-bus system, which does not require costly and heavy switching devices, with 5,000 separate cables weighing 490 tons and a cooling system of 400 tons.

The SSPS must keep its solar arrays oriented toward the Sun. This will require an argon reaction mass of 15–50 tons per year in addition to that required for station-keeping. Both maneuvers will produce an interaction between the solar arrays and the satellite body. This dynamic response problem must be solved as well as that of the oscillation of the SSPS over an area of about 50–150 km^2. This latter problem may be crucial to the exact pointing of the Earth satellite CETI radiotelescope. Also, it is important to control the interference to the CETI radiotelescope by the SSPS, especially for passive CETI where the latter's very high sensitivity must be taken into account. Another source of powerful radio frequency interference stems from the normal electromagnetic activity of our civilization (radio, TV, etc.). A possible solution is to screen the CETI radiotelescope from both the Earth and the SSPS.

References

Dixon R. S. (1973) A search strategy for finding extraterrestrial radio beacons. *Icarus* **20**, 198–199.

Dixon R. S. and Cole D. M. (1977) A modest all-sky search for narrow-band radio radiation near the 21-cm hydrogen line. *Icarus* **30**, 267–273.

Gerritsen H. J. and McKenna S. J. (1975) The Luneberg lens and the importance of transmission in establishing contact with extraterrestrial civilizations. *Icarus* **26**, 250–256.

Gohrbandt B. and Schmidt E. F. (1976) *Space power stations—an analysis of problem.* Paper presented at the IAF Congress in Anaheim, California.

Oliver B. M. and Billingham J. (Directors) (1973, revised edition) *Project Cyclops: a Design Study of a System for Detecting Extra-terrestrial Intelligent Life.* NASA CR 114445.

Subotowicz M. (1976) Solar satellite power stations and satellite energy re-transmitters (in Polish), *Postepy Astronautyki* (Progress in Astronautics), Warsaw 9, 4–47, Nr 3/26.

Subotowicz M. (1978) *CETI from an Earth satellite orbit,* paper presented at an IAF Congress in Anaheim, California, 1976, and published in JBIS Interstellar Studies Issue (1978), **31**, 109–110, No. 3.

Acta Astronautica. Vol. 6, pp. 213–220. Pergamon Press 1979. Printed in Great Britain

Interstellar communication by neutrino beams

M. SUBOTOWICZ†

Institute of Physics, University M. Curie-Sklodowska, Lublin, Poland

Abstract—The general foundations concerning the physical, chemical, and astronomical conditions for the existence of life are presented. The properties of the neutrino are described and the reactions that produce and detect them are given. Basic problems in neutrino physics are listed and the possibility of producing and detecting large anisotropic neutrino beams is discussed. Two possibilities are considered: anisotropic beams from accelerators of charged particles, and from the polarized beta-radioactive nuclei. Neutrino beam fluxes for different distances, 1000 and 10 light years, and for different beam characteristics are calculated. The problem of detecting neutrino beams from such distances is discussed.

1. Introduction

THE DIAMETER of the Universe as we observe it is about 10^{10} light years. There are about 10^{10} galaxies in this part of the Universe, or about 10^{20} stars. The chemical composition of matter is the same everywhere in the Universe, and the same physical laws rule as well. Astronomers accept now that the appearance of planetary systems is one of the steps in the formation of stars, although this has not yet been proved by observation. Among the 10^{20} stars, there should exist a sufficiency of planetary systems. On some of the planets, physical and chemical conditions should be similar to those on Earth. We assume that the appearance of life on these planets is a normal process in the evolutionary development of matter (chemical and biological evolution), including intelligent beings (Cameron, 1963).

We assume also that the scientific-technical civilization (STC) developed here on Earth by homo sapiens is typical and common, and that its present scientific-technical level can be achieved—on the average—after the same elapsed time from the beginning of evolution (Oparin, 1953, 1961; Fox, 1965). We can hope also that there exist in the Universe other STC's such as ours, or even some much more advanced than here on Earth.

We can define an STC by the particular stage of development in radioelectronics, achieving a level sufficient (a) to transmit powerful signals that could be received and recognized by the nearest STC at least as advanced as we are and (b) to receive signals of minimal power transmitted by the nearest possible STC (von Hoerner, 1961; Cocconi and Morrison, 1959; Schwartz and Townes, 1961).

As a very rough estimate, only one STC can develop, on the average, for every 10^6 to 10^{12} stars in the Universe. The details of this calculation are not given here; e.g., von Hoerner (1961). Consequently, the average distance between two STC can be from hundreds to many thousands of light years. This

†Professor Doctor, Head of Department of Experimental Physics.

distance in the worst case is comparable to the dimensions of the Galaxy. Thus, generally, the problem of communication with extraterrestrial STC involves the following topics: (a) the existence of STC's and their development from the astronomical, chemical, and physical points of view; (b) the development of intelligent beings by chemical and biological evolution up to the creation of the STC's; and (c) the discovery of and communication with the extraterrestrial STC's by our terrestrial STC (Vnezjemnyje Tsyvilisatsji, 1965; Fox, 1965).

In this paper, only the last question is considered. The possibility of interstellar communication by use of electromagnetic signals has been discussed by many authors (Kaplan, 1969; Dyson, 1960; Oliver and Billingham, 1973; Subotowicz, 1973a,b). I would like to discuss now another possibility: sending messages over interstellar distances using neutrino beams.

2. Neutrino beams

The neutrino is one of the most mysterious of elementary particles (Kuchowicz, 1966; Jatczak, 1965). It is a massless lepton, its spin is $\pm 1/2$, and it probably has a magnetic moment. Its interaction with matter is extremely weak, the cross section for this interaction being about 10^{-43} cm^2. The incoming neutrino flux reaches us from its origin without any changes in its intensity, direction, or energy. Therefore, a neutrino would be an ideal medium for interstellar communication. But there are certain problems: (a) How can one produce very intense neutrino beams of high directivity? Neutrino signals should be distinguishable from solar neutrinos, 10^{10} ν/cm^2-s, and from the cosmic and atmospheric background neutrinos, 10^{-7} to 10^{-3} ν/cm^2-s; (b) How can one detect neutrino beams reaching the consignee from another STC?

3. Production and detection of neutrinos

The following elementary processes are known to generate a neutrino.

(a) Production of neutrinos in the presence of mesons, nucleons, or nuclei:

$$n \rightarrow p + e^- + \bar{\nu}$$

$$e^- + p \rightarrow n + \nu \text{ (Reines and Cowan, 1953)}$$

$$^3\text{He} + e^- \rightarrow {}^3\text{H} + \nu$$

$$^3\text{He} + p \rightarrow {}^4\text{He} + e^+ + \nu$$

$$p + p \rightarrow d + e^+ + \nu$$

$$p + p + e^- \rightarrow d + \nu$$

URCA process ($\rho > 10^5$ g/cm^3, $T > 10^8$ °K):

$${}^A_Z\text{N} + e^- \rightarrow {}_{Z-1}^{\ \ A}\text{N} + \nu$$

$${}_{Z-1}^{\ \ A}\text{N} \rightarrow {}^A_Z\text{N} + e^- + \nu$$

$$\pi^\pm \rightarrow \mu^\pm + \nu \quad (\text{or } \bar{\nu})$$

$$K^{\pm} \rightarrow e^{\pm} + \nu \quad (\text{or } \bar{\nu})$$

$$e^{-} + {}_{Z}N \rightarrow e^{-} + {}_{Z}N + \nu + \bar{\nu}.$$

(b) Pure lepton processes:

$$e^{-} + e^{+} \rightarrow \nu + \bar{\nu}$$

$$e^{+} + e^{-} \rightarrow e^{+} + \mu^{-} + \nu + \bar{\nu} \quad (\text{if} \quad \bar{\nu}_{\mu} = \bar{\nu}_{e})$$

$$\mu^{+} + e^{-} \rightarrow \nu + \nu^{-}$$

$$e^{-}_{\text{free}} \rightarrow e^{-}_{\text{bounded}} + \nu + \bar{\nu}.$$

(c) Photon processes:

$$\gamma + \gamma \rightarrow \nu + \bar{\nu} \quad \text{or} \quad \gamma + \gamma \rightarrow \gamma + \nu + \bar{\nu}$$

$$\gamma + N \rightarrow N + \nu + \bar{\nu}$$

$$\gamma + e^{-} + N \rightarrow N + e^{-} + \nu + \bar{\nu}$$

$$\gamma + e^{-} \rightarrow e^{-} + \nu + \bar{\nu}$$

$$\gamma + p \rightarrow n + e^{+} + \nu$$

$$\gamma + n \rightarrow p + e^{-} + \bar{\nu}$$

(d) Plasma processes:

A neutrino particle is extremely difficult to detect; this is the obvious disadvantage in the present context. The cross section for neutrino interaction measured in the well-known Reines–Cowan (1953) experiment was $\sigma \sim 10^{-43}$ cm^2. For neutrino energy (in CGS units) approaching infinity, the cross section (Reines, 1961) is $\sigma \rightarrow 10^{-38}$ cm^2, which is also very low.

Bubble chambers, Čerenkov counters, liquid scintillation counters, and spark chambers have been used as detectors in neutrino experiments. In the low-energy region, detection of neutrinos with energies up to several MeV has been based on:

(a) Inverse β-decay (Reines and Cowan, 1953):

$$\bar{\nu}_{e} + p \rightarrow e^{+} + n \quad (\sigma \sim 10^{-43}\,\text{cm}^2).$$

(b) Davis radiochemical experiment (Davis, 1955):

$${}^{37}\text{Cl} + \nu_{e} \rightarrow {}^{37}\text{Ar} + e^{-}.$$

(c) Elastic neutrino scattering:

$$\bar{\nu} + e^{-} \rightarrow \bar{\nu} + e^{-} \quad (\sigma \sim 5.10^{-45}\,\text{cm}^2).$$

In the high-energy region, neutrinos can be detected in the following ways:

(a) Brookhaven-type experiment (Danby *et al.*, 1962) ($\sigma \sim 10^{-38}\,\text{cm}^2$):

$$\nu_\mu + n \to p + \mu^- \quad \text{or} \quad \bar{\nu}_\mu + p \to n + \mu^+$$
$$\nu_e + n \to p + e^- \quad \text{or} \quad \bar{\nu}_e + p \to n + e^+.$$

(b) CERN-type experiment (Bernardini, 1964) to detect the intermediate boson W:

$$\nu_\mu + \text{N} \to W^+ + \mu^- + \text{N}$$
$$W^+ \to \mu^+ + \nu_\mu \quad \text{or} \quad W^+ \to e^+ + \mu_e \quad \text{or} \quad W^+ \to \begin{cases} \text{kaons} \\ \text{pions} \end{cases}.$$

(c) Johannesburg-type experiment (Reines *et al.*, 1965):

$$\nu_\mu + \text{N} \to \mu + \text{N}'.$$

(d) Elastic neutrino-nucleon scattering:

$$\nu + p \to \nu + p \quad \text{and} \quad \bar{\nu} + n \to \bar{\nu} + n.$$

(e) Neutrino-electron scattering:

$$\bar{\nu} + e^- \to \bar{\nu} + e^-, \quad \nu + e^- \to \nu + e^-.$$

(f) Reaction with deuterons:

$$\bar{\nu} + d \to e^+ + n + n.$$

4. Neutrinos in interstellar and global communication

Independent of the idea of neutrino communication proposed here, many important basic problems of the physics of weak and other principle interactions, as well as problems in astrophysics and cosmology, must be solved in the area of neutrino physics. Therefore, any experimental progress in solving the basic scientific problems of neutrino physics could accelerate the use of neutrino beams for interstellar communication.

4.1. *Basic problems in neutrino physics*

- Lepton conservation law in the region of large neutrino energies
- Muon and electron neutrinos
- Existence and role of the intermediate boson in weak interactions
- Existence of a background of degenerate neutrinos (and antineutrinos), and the Fermi level of the neutrino gas: does matter convert into neutrinos?
- Energy spectrum of cosmic neutrinos
- Neutrino neutral current

4.2. *Basic neutrino problems in astrophysics and cosmology*

- Neutrino telescope
- Role of neutrinos in the energy balance of stars
- Neutrino and electromagnetic radiation luminosity of stars
- High-temperature and high-pressure neutrino processes in the non-equilibrium state of a star
- Stars made of antimatter
- Cosmogonic history of the Universe and the presence of mesonic and electronic neutrinos
- Nuclear processes in the interior of stars, especially that of the Sun, investigated by the neutrino spectrum method; solar neutrino puzzle
- Average density of neutrino matter in the Universe, and the density of the nonzero rest-mass matter
- Chemical composition of stars, and neutrino processes in stars
- Universe models and neutrino energy density

4.3. *Neutrino in interstellar and global communication*

Because of its very weak interaction with interstellar and atmospheric matter, the neutrino could be taken by representatives of an advanced STC as the best possible and most reliable medium for sending signals and information to other STC's. Therefore, a definition of an STC should contain instructions on how to send and receive information by means of a neutrino beam. The frequency or, better, the energy-modulated neutrino signals sent by the extraterrestrial intelligence could carry the desired information. A more precise definition of the STC should present the conditions for sending and detecting proper neutrino signals. Monochromatic neutrino beams could serve as the identification signals. For high-energy neutrinos (about 1 GeV), beams could be obtained with accelerators having a large number of protons in each pulse. For low-energy neutrinos (< 15 MeV), the sender should transmit signals that are distinguishable from the mother-star background signals. Therefore, the neutrinos should be sent anisotropically. It is well known that, because of parity violation in weak interactions, the β-electron and neutrino emission of polarized radioactive atomic nuclei is highly anisotropic. Single crystals comprised of the proper radioactive nuclei (atoms), which can be polarized by a magnetic field when at a very low temperature, will emit such anisotropic neutrino radiation from the decaying nuclei.

4.4. *Anisotropic neutrino beams from accelerators*

The key problem in neutrino communication is directivity of the radiation, because the receiver should register an intensity above that of the background. As an artificial source of anisotropic neutrino radiation, accelerators of charged particles (protons) can be used. The particles that bombard the target nuclei will generate $\pi^{\pm}$, $K^{\pm}$ mesons; those that decay will emit μ^{-} mesons, electrons, and $\nu,\bar{\nu}$ in a narrow and well-collimated beam. The accelerator method for producing neutrinos is summarized as follows:

$$p + \mathrm{N} \rightarrow \pi^+ + \pi^- + \mathrm{N}'$$

$$\pi^- \rightarrow \mu^- + \bar{\nu}_\mu \quad \text{or} \quad \pi^- \rightarrow \bar{\nu}_e + e^-,$$

$$\pi^+ \rightarrow \mu^+ + \nu_\mu \quad \text{or} \quad \pi^+ \rightarrow \nu_e + e^+ \quad \text{(pion decay)}$$

$$\mu^+ \rightarrow \bar{\nu}_\mu + \nu_e + e^+$$

$$\mu^- \rightarrow \nu_\mu + \bar{\nu}_e + e^- \quad \text{(muon decay).}$$

Let the energy of the protons be from several GeV to 25 GeV. Then the energy of the pions in the beam will be from about 0.5 GeV to several GeV. For a well-collimated beam of pions (about 1°), their momentum will be several GeV/c. In this case, the forward direction of the neutrino emission will occur in a cone with an apex angle of about 3°.

Suppose that an accelerator pulse contains 10^{15} protons and that there are 10^3 pulses/hr. In Table 1, the experiments assumed are much more advanced than

Table 1. Neutrino (or antineutrino) flux from an accelerator, for different energies of the primary proton beam, for different beam directivities, and for different distances from the neutrino source

Proton energy, GeV	Distance, light years	Neutrino flux F (ν/cm^2-hr)		
		Beam	Directivity	
25 (each proton produces 10 pions)		3°	20′	
	1000	5.2×10^{-21}	5.2×10^{-19}	
	10	5.2×10^{-17}	5.2×10^{-15}	
300 (each proton produces 100 pions)		1°	1′	1″
	1000	7.5×10^{-19}	2×10^{-16}	7×10^{-13}
	10	7.5×10^{-15}	2×10^{-12}	7×10^{-9}

are possible now: accelerators for 300 GeV; proton pulses on the order of 10^{15}; beam collimation up to 1′ or even 1″; and a detection sensitivity for fluxes in the range 7×10^{-9} to 5×10^{-21} ν/cm^2-hr. Up to now, the weakest experimentally detected neutrino beams were on the order of 3×10^{-3} ν/cm^2-s—the result of primary-cosmic-ray interaction with the nuclei of the atoms in the atmosphere. This flux is 10^6 times larger than the largest flux in Table 1.

4.5. *Neutrinos from polarized radioactive nuclei*

The second possibility we see is the emission from polarized radioactive nuclei. The directivity of the neutrino beam is a function of the polarization of the radioactive nuclei and the neutrinos from the β-decay have energies up to

Table 2. Neutrino (or antineutrino) flux from the polarized ^{60}Co radioactive nuclei for different beam directivities and different distances from the beam source

Average neutrino energy, keV	Distance, light years	Neutrino flux F (ν/cm^2-hr)	
		Neutrino beam directivity	
		1′	1″
Several	1000	8.6×10^{-12}	2.9×10^{-8}
hundred	10	8.6×10^{-8}	2.9×10^{-4}

several MeV. For a great quantity of ^{60}Co atoms (about $1\ m^3$), we get a directivity of 1′ and fluxes from 10^{-8} to 10^{-12} ν/cm^2-hr, which means beams four to five orders of magnitude larger than for the accelerator neutrinos (as shown in Table 2). But for neutrino energies of several MeV or less, the cross sections are four to five orders of magnitude smaller than for high-energy neutrinos (~ 1 GeV). Another problem is how to polarize the β-radioactive nuclei of $1\ m^3$ of ^{60}Co so that the neutrino beam directivity is about 1 min of arc or even 1 sec of arc.

5. Conclusions

It is as difficult for interstellar communication to generate and detect neutrinos from accelerators as from polarized nuclei. Contemporary techniques for detecting neutrinos do not allow a determination of the direction of their origin. We now distinguish neutrinos according to the threshold energy of the particular reaction that serves as the detection process. The basic difficulties encountered are: the high level of the background, the large dimensions of the detectors, and their very low detection efficiency. This is not the case when using neutrinos for telecommunication over global distances, 10^3–10^4 km through the Earth.

There are major difficulties in solving the problem: how to generate well-collimated neutrino beams and how to detect them at distances from several light years up to perhaps 1000 light years.

6. Remarks added in proof (1978)

6.1 In principle, besides electromagnetic waves and neutrino beams, there should be a third medium through which one might achieve CETI—gravitational waves: But until now they have not been detected (Pizzella, 1978). Therefore, in 1966, when the main ideas of the present paper were published (Subotowicz, 1967), I regarded any serious discussion of this CETI variant as premature. In this respect I have not so far changed my opinion.

6.2 In the recent paper by Sáenz *et al.* (1977), neutrino beams in the 1–100 GeV energy range, from high energy proton accelerators, are proposed as potential means for telecommunication over global distances, 10^3–10^4 km. Neutrino beams propagating through the Earth for such distances will not be attenuated or scattered to an appreciable degree. Similar proposals for telecommunication not only over global but also interstellar and intergalactic distances were investigated in a decade earlier paper (Subotowicz, 1967).

6.3 It is not excluded that some advanced civilizations may deliberately shutout emergent civilizations such as ours from the conversation. This implies the direct antithesis of the anticryptographic, simple-talk hypothesis. Generation and detection of neutrinos is so difficult that an advanced civilization may purposely choose such a system in order to find and communicate only with STCs at their own level of development.

6.4 Recently there have been proposed new and more efficient methods for neutrino detection; e.g. the acoustic method, with the volume of the detecting liquid measured in cubic kilometers of ocean water!

References

Bernardini G. (1964) Some motivations for a very costly long-range programme on high-energy neutrinos. *Proc. Roy. Soc.* (London) **278A**, 369–390.

Cameron A. G. W. (Editor) (1963) *Interstellar Communication.* New York.

Cocconi G. and Morrison P. (1959) Searching for interstellar communications. *Nature* **184**, 844–846.

Danby G., Gaillard J. M., Goulianos K., Lederman L. M., Mistry N., Schwartz M. and Steinberger J. (1962) Observation of high energy neutrino reactions and the existence of two kinds of neutrinos. *Phys. Rev. Letters* **9**, 36–44.

Davis R. Jr. (1955) Attempt to detect the neutrino from a nuclear reactor by the $^{37}Cl(\bar{\nu},e^-)\ ^{37}A$ reaction. *Phys. Rev.* **97**, 766–769.

Dyson F. G. (1960) Search for artificial stellar source of infrared radiation. *Science* **131**, 1667–1668.

Fox S. W. (1965) *The Origins of Prebiological Systems.* Academic Press, New York.

Jatczak J. (1965) Astrofizyka neutrinowa (in Polish) *Postepy Astron.* **13**, 63–84.

Kaplan S. A. (Editor) (1969) *Vnjezjemnyje tsivilisatsji* (in Russian). Nauka, Moscow.

Kuchowicz B. (1966) *The bibliography of the neutrino, I and II.* Nuclear Energy Information Center, Warsaw.

Oliver B. M. and Billingham J. (Directors) (1973), Revised edition *Project Cyclops: a Design Study of a System for Detecting Extratterrestrial Intelligent Life.* NASA CR 114445.

Oparin A. J. (1953, 1957) *The Origin of Life.* Moscow.

Oparin A. J. (1961) *Life: Its Origin. Nature and Development*, London.

Pizzella G. (1978) Recent research on gravitational waves. *Europhysics News* **9**(9), 4–8.

Reines F. (1961) Neutrino interactions. *Ann. Rev. Nuclear Sci.* **10**, 1–26.

Reines F. and Cowan C. L. (1953) The detection of the free neutrino. *Phys. Rev.* **92**, 8301.

Reines F., Crouch M. F., Jenkins T. L., Kropp W. R., Gurr H. S., Smith G. R., Sellshop J. P. F. and Mayer B. (1965) Evidence for high-energy cosmic-ray neutrino interactions. *Phys. Rev. Letters* **15**, 429–433.

Sáenz A. W., Überall H., Kelly F. J., Padgett D. W. and Seeman N. (1977) Telecommunication with neutrino beams. *Science* **198**(4314), 295–297.

Schwartz R. N. and Townes C. H. (1961) Interstellar and interplanetary communication by optical masers. *Nature* **190**, 205–208.

Subotowicz M. (1967) Neutrina w lacznosci miedzygwiezdnej (in Polish). *Post. Tech. Jadr.* **11**, 475–479.

Subotowicz M. (1967) Interstellar communication with electromagnetic and neutrino signals (in Polish). *Astronautyka* **10**, 19–24.

Subotowicz M. (1973a) W poszukiwaniu pozaziemskich cywilizacji: przeslanki, propozycje i cele programu "Cyklop" (in Polish). *Astronautyka* **4**, 2–4.

Subotowicz M. (1973b) W poszukiwaniu pozaziemskich cywilizacji: program "Cyklop" nawiazania lacznosci miedzygwiezdnej (in Polish). *Astronautyka* **5**, 2–6.

von Hoerner S. (1961) The search for signals from other civilizations. *Science* **134**, 1839–1843.

Vnezjemnyje Tsyvilisatsji (1965) *Proc. Conf. Interstellar Communication*, Byurakan.

Acta Astronautica. Vol. 6, pp. 221–225. Pergamon Press 1979. Printed in Great Britain

Message through time

GEORGE MARX

Department of Atomic Physics, Roland Eötvös University, Budapest, Hungary

Abstract—If one takes the time difference among galactic civilizations into account, small unmanned space probes of long duration have been considered as a possible channel for interstellar communication. Criticizing the electronic engineering approach, we offer some nonconformist suggestions which make use of the possibilities offered by different sciences.

Electronic probes

AFTER THE pioneering idea of Tesla, on the basis of a solid scientific analysis, Bracewell (1960) concluded, first, that interstellar space probes might be economically viable for medium-range interstellar communication. An automatic computerized radio station in a circumplanetary orbit at a target planet could wait for the emergence of a technological civilization, thereby establishing the first radio contact with the developing race. This idea was popularized by Lunan (1973). He claimed that the delayed radio echos, observed several times in our century, might contain messages relayed by such a probe, which had been put in orbit at the Trojan point of the Earth–Moon system a few thousand years ago (Strong, 1965). The physical existence of unexplained delayed echos is a reality, but their interpretation as intelligent messages is highly dubious (Lawton, 1973, 1974).

The "radio transmitting probe" as a technical solution for the distance–time problem of interstellar contact may attract basic criticism. A time gap of 10,000 years between the arrival of the probe and the emergence of the terrestrial civilization seems to be a surprisingly narrow one. The time separation between the birthdates of intelligent races may be as great as several millions—even billions—of years. Knowing this, one cannot base the search for contact only on the assumption of the simultaneity between transmission and reception. An expanding intelligent race might be interested in sending information through space and time to another planetary system and to a much later era. But the automatics of our interplanetary probes do not survive more than one or two years. Time is a harder obstacle for electronics than distance.

Nuclear probes

Most atomic nuclei are stable formations everywhere in the Universe, except in the central regions of the hottest stars. The isotopic composition of material samples is not modified either by fire or by slow corrosion. This fact offers a possibility for writing a highly durable letter: the proton and neutron numbers in a piece of solid material may be used as letters to be read by mass spectroscopy. The charge number of the elements could give the ordering, and the deviation from the most stable mass number could give the character. In this way even a tiny pebble of artificial isotopic composition might contain a message which carries more than 100 bits of information and which is highly resistant to chemical corrosion.

But this information is no longer than a short sentence. One can enrich the information content of the message by using the charge and mass number for ordering and the isotopic abundances as characters. Three-hundred stable isotopes with 1% accurate concentration values may carry $300 \log_2 100 \approx 2000$-bits information, which is equivalent to a 400-letter telegram. A tiny piece of alloy, first molten, then resolidified with anomalous isotopic composition, may hide a letter that was written millions of years ago. It could contain ten times more information than Lunan's "star maps" constructed from the delayed radio echos.

The main difficulty with such nuclear letters is that they can be lost easily. It has been discovered that some recent meteorites show peculiar isotopic abundances, but most prehistoric meteorites are forever lost. The problem of locating such prehistoric letters appeals to the fantasy of science-fiction writers (Clarke, 1963).

Biological probes

Spontaneous generation of life does not occur on the Earth nowadays. The most popular hypothesis for the origin of terrestrial life is that it was born on Earth just after the formation of the planet when atmospheric conditions were more favorable from a chemical point of view, maybe within the first 100 million years after the appearance of the fluid oceans. Nothing definite is known about the probability of such a spontaneous birth. The formation of amino acid molecules has been observed in outer space and in the laboratory, but the first living beings had to be highly sophisticated systems capable of carbon and nitrogen metabolism under inorganic conditions. They might not have been much simpler than the blue–green algae. Their spontaneous formation may have an extremely low probability. Arrhenius (1908), Oparin (1957), and others suggested the idea of panspermia: Earth had been infected by life from outside. An objection is that radiation would destroy any free spores in outer space. The arrival at Earth of a meteorite from another planetary system is not likely to happen.

Seeing these difficulties, Crick and Orgel (1973) suggested the idea of directed panspermia: Earth has been infected by intention. An artificial space probe may have brought to Earth deep-frozen microorganisms which became the seeds of terrestrial evolution. According to Crick and Orgel, a faint indication suggesting an extraterrestrial origin of terrestrial life is the universality of the terrestrial genetic code and the surprising importance of molybdenum in some enzymes. Molybdenum is a very rare element in the Solar System (see accompanying table), but one must admit that it is a very efficient coordinating center for CN groups in complex organic molecules. The surprisingly fast emergence of a rather developed life form after the formation of oceans (Orgel, 1973) could be explained by this assumption of directed panspermia.

Table 1. Relative Solar System abundances of some elements essential to life on Earth.

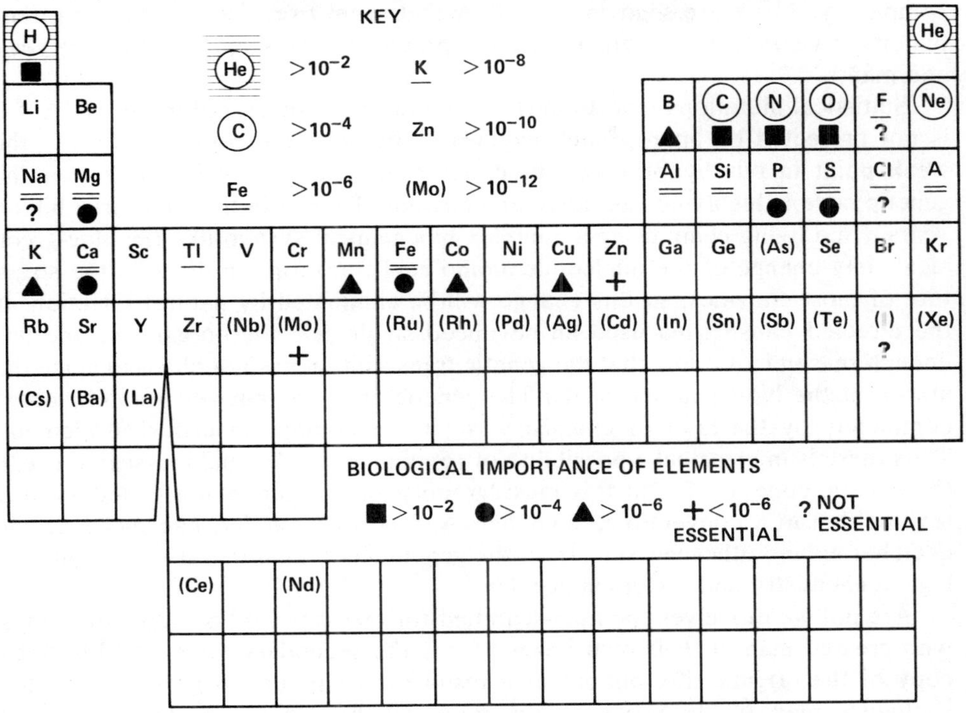

Our simplest self-supporting life forms have a DNA chain with 5×10^6 nucleotides, which means 5×10^6 letters in a four-character alphabet. This means $5{,}000{,}000 \log_2 4 \simeq 10^7$ bits, equivalent to a normal book with 1000 pages. It is a well-known biological observation that a considerable fraction of this "genetic" information is apparently inactive from a biological point of view. (The biologically inactive base sequence is partly self-repeating, but partly "random", carrying some sort of information.)

How does one send a letter to a faraway planet, a letter that is light enough for easy transportation, that multiplies itself on arrival, that can correct misprints automatically, and that will be read definitely by the intelligent race of the target planet after they have reached scientific maturity?

Sending information-carrying and self-reproducing chain molecules is perhaps one of the most economical solutions. It is possible that a few billion years ago an advanced civilization prepared some sort of message using genetic engineering and sent it to Earth. This extraterrestrial DNA molecule became the starting point of biological evolution. It might be that the message was only: "life is possible". On the other hand, DNA is large enough to contain both biochemical recipes and written intellectual messages. Our intelligent human civilization, emerging from this seed, will perhaps read the message written by our parents in our own nucleotide sequence (to be deciphered in the near future). Perhaps the mutations—necessary for evolution—have not yet destroyed the original text completely. There are segments of DNA that seem to resist mutations through billions of years. (The protein histone has practically the same structure for yeast and man.)

Statistical mutations tend to destroy any information stored in the DNA if it is not protected by appropriate enzymes or by selection pressure. This is the weak point in our hypothesis. But it has been stressed by F. Crick that the genetic code is highly conservative and has not changed in the past three billion years. (In a living being using a complex biochemical system like the blue-green algae, any change of the nucleotide-amino acid transcription disturbs the structure of most enzymes, so this change will be eliminated by natural selection. If the directed panspermia used an advanced single-cell microorganism, one has enough reason to assume that the genetic transcription has not changed since the arrival of the biological message.) The genetic code has not yet been explained completely by the laws of chemistry, so it can contain an historical element. Thus there is in principal a possibility of sending an intellectual message through the genetic code itself. But this message must be shorter than a 40-letters long text written in a conventional alphabet. And nobody, so far, has succeeded in deciphering any other message from the genetic code than that it is a mixture of logical chemistry and historical accident.

Are not we ourselves the long-searched-for message sent by superior beings, who created man in their own image? Or is the secondary terrestrial life not a copy of the original life, but rather a result of computerized genetic planning? Biologists may try to prove or disprove this hypothesis. Is the abundance distribution of biological trace elements a chemical necessity or does it reflect the composition of a terrestrial puddle, or the chemistry of the birth planet of life? Does a part of the inactive DNA segments carry information? Is this segment similar in undeveloped and developed life forms? Can the sudden appearance of perfect organisms be explained in this way—a nose that is sensitive to one molecule, an eye sensitive to one single photon, a human brain with a capacity not yet exploited completely? All this seems to hint at an engineering perfection.

The aim of this paper was not to answer these questions. I intended only to note the need for a more open-minded approach to the problem of interstellar communication through space and time. The preference for electromagnetic signals may be only a prejudice of our television-oriented twentieth century. The Viking mission to Mars raises questions about the origin of life in our Solar System.

The author is indebted to Mr. A. C. Clarke and Prof. V. Csányi for valuable discussions.

References

Arrhenius A. (1908) *Worlds in Making.* Harper and Row, New York.
Bracewell R. N. (1960) Communications from Superior Galactic Communities. *Nature* **186**, 670.
Clarke A. C. (1963) 2001, *a Space Odyssey.* Scott Meredith, New York.
Crick F. H. C. and Orgel L. E. (1973) Directed Panspermia. *Icarus* **19**, 341.

Lawton A. T. (1973) The Interpretation of Signals from Space. *Spaceflight* **15**, 132.
Lawton A. T. (1974) Interpreting Signals from an Interstellar Probe. Presented at XXVth Congress of IAF, Amsterdam.
Lunan D. A. (1973) Space Probe from Epsilon Bootis. *Spaceflight* **15**, 122.
Oparin A. (1957) *The Origins of Life on Earth.* Academic Press, New York.
Orgel L. E. (1973) *The Origins of Life.* John Wiley, New York.
Strong J. (1965) *Flight to the Stars.* Temple Press, London.

ACTA ASTRONAUTICA

Journal of the International Academy of Astronautics

Vol. 6 No. 1–2 Jan.–Feb. 1979

Ex Mundo Astronautico

CONTENTS

TABLE DES MATIÈRES

СОДЕРЖАНИЕ

Acta Astronautica. Vol. 6, pp. 3–12. Pergamon Press 1979. Printed in Great Britain

Ex Mundo Astronautico

International Academy of Astronautics Report of the President for the Year 1977–1978

Professor C. S. Draper (*President*)

(*Report prepared* 31 *August* 1978)

1. Introduction

THE International Academy of Astronautics, now eighteen year old, will soon complete the second decade of its existence. This may be the time to reconsider the tremendous, rapid advances that have been made in the development and use of space technology, and to look beyond the technical achievements to other areas in which they have an indubitable impact.

Much thought is being given to the humanistic, psychological, and philosophical factors that are involved in the utilization of space technology and will become more important as man proceeds further into out space. This trend is already reflected in the choice of the theme for the 29th Congress of the International Astronautical Federation, viz. "Astronautics for Peace and Human Progress".

It seems to me that the Academy cannot remain aloof from these considerations and will have to develop activities that take into account the past achievements and the new roads opening up to the future. The next Regular Meeting of the Academy, which will be held in 1979, should reveal interesting projects in this direction.

The 9th Regular Meeting of the Academy took place in Prague on 29 September 1977. On-going activities were reviewed and discussed, as well as possible new areas of study. The current financial situation and ways to increase the Academy's resources were also discussed.

During the meeting the following officers were elected for two-year terms:

President:	C. S. Draper (U.S.A.)	
Vice-Presidents:	H. A. Bjurstedt (Sweden)	A. Mikhailov (U.S.S.R.)
	E. A. Brun (France)	L. G. Napolitano (Italy)

Two Trustees of each Section whose terms expired in 1977 were re-elected for another two years.

The Board of Trustees met in Prague in September 1977 and again in Paris in 1978, and gave much attention to the financial problems. No concrete solution has been found yet, although various suggestions put before the Board are being examined by the Finance Committee. Some conclusions and decisions are expected to be reached during the meeting that will be held in Dubrovnik this coming October.

The Finance Committee for the next two years was appointed by the Board of Trustees as follows:

Chairman: R. Pesek (Czechoslovakia)
Members: A. Jaumotte (Belgium) and L. G. Napolitano (Italy)

Publications were also a subject of concern for the Board of Trustees, particularly as regards *Acta Astronautica.* An *ad hoc* Committee of Board members was designated to make a study of all aspects of the policies and production of the journal.

In Prague, the Board of Trustees also appointed the Award and Fellowship Committee for the next two years:

Chairman:	C. S. Draper (U.S.A.)	
Vice-Chairman:	M. Nicolet (Belgium)	
Members:	C. A. Berry (U.S.A.)	M. V. Keldysh (U.S.S.R.)
	H. A. Bjurstedt (Sweden)	R. Lust (F.R.G.)
	L. Broglio (Italy)	L. I. Sedov (U.S.S.R.)

Unfortunately, the Committee has since lost one of its members, Academician M. V. Keldysh, who died in June 1978.

2. Membership

At the 9th Regular Meeting of the Academy, an increase in the quota for Members in all Sections was voted, thereby creating additional vacancies for the December election. As a result, 21 Members were elected and 25 new Corresponding Members were admitted to the Academy.

The Academy regrets the loss of the following members through death:

August 1977	—Howard S. Seifert, Member, Section 2, since 1960
September 1977	—A. V. Cleaver, Founding Member, Section 2
March 1978	—André Lallemand, Corresponding Member, Section 1, since 1967
May 1978	—Harry E. Ross, Corresponding Member, Section 2, since 1967
May 1978	—Fernand Vinsonneau, Corresponding Member, Section 2, 1965; Member since 1971
June 1978	—Mstislav V. Keldysh, Corresponding Member, Section 2, 1965; Member since 1965
July 1978	—Henri Moureu, Corresponding Member, Section 2, 1965; Member since 1971

The present membership of the Academy is composed as follows, making a total of 540:

Honorary Members: 11	
Section 1, Basic Sciences:	70 Members— 87 Corresponding Members
Section 2, Engineering Sciences:	76 Members—122 Corresponding Members
Section 3, Life Sciences:	70 Members—104 Corresponding Members

A list of the membership as of August 1978 is being sent to all members; additional copies are available upon request from the Secretatiat.

3. Academy Activities

A. *Awards*

The 1977 Daniel and Florence Guggenheim International Astronautics Award was announced and presented to Prof. L. I. Sedov (U.S.S.R.) during the 28th I.A.F. Congress in Prague last September. The Award was made in recognition of Professor Sedov's recent significant contributions to the mathematical theory of inertial navigation based on relativistic effects, as well as his well known work in the fields of navigation and continuum mechanics, such as similarity laws, turbulent flows, gasdynamics, and his recent modelling of systems with a rational approach to these problems.

The recipient of the 1978 Daniel and Florence Guggenheim International Astronautics Award is Dr. Christopher C. Kraft, Jr., who has directed the planning and operational control of the U.S. manned space flight programs from the first suborbital Mercury flight through to the current development of the Space Shuttle Program. Dr. Kraft has been Director of the NASA Johnson Space Center, at Houston, Texas, since 1972. His knowledge and experience in engineering and other scientific disciplines contributed appreciably to the definition and design of the Space Shuttle Program, and the development of this new concept of space transportation in general. Another outstanding contribution to manned space flight was his early perception of the need to establish alternate courses of action and procedures in case of system failures or other contingencies during flight.

The presentation of the 1978 Award will take place during the 29th I.A.F. Congress in Dubrovnik in October.

B. *Scientific Meetings*

1. *International Space Rescue and Safety Symposia*

(a) The 10th International Space Rescue and Safety Symposium, consisting of two sessions, was held in Prague on 30 September and 1 October 1977. The sessions were chaired respectively by P. H. Bolger (U.S.A.) and J. Dvorak (Czechoslovakia). Details of the programme are given in Annex I, para. 1.

As a complement to the Symposium, a panel discussion was held on 29 September, under the title: "Worldwide Disaster and Rescue Programs". The discussion was organized and conducted by G. W. Heath (U.S.A.). Panelists representing various international and national organizations, governmental as well as non-governmental, discussed experimental space-borne systems and others under development for land, sea, or air disaster/rescue response practice. Among the factors cited were limitations in rescue force equipment, personnel capabilities, and terrain/environmental limitations; jurisdictional gaps or overlappings that stand in the way of swift, effective action being taken at a disaster/rescue site and the difficulties of changing previously organized response systems. A realization of the potential of applying space technology to rescue/relief operations was considered essential to meet the challenges.

(b) The 11th Symposium of this series, under the general chairmanship of P. H. Bolger, will be held during the 29th I.A.F. Congress in Dubrovnik, next October. One session on Space Flight Safety will be chaired by J. Hammock (U.S.A.) and will consider in-flight risks and their solutions. A second session, organized by G. W. Heath, will deal with Earth Rescue and Recovery Safety. The programme includes reports by invited speakers from six different countries: Canada, F.R.G., Indonesia, Norway, Switzerland, U.S.A., with emphasis on disaster response requirements and systems.

2. *International History of Astronautics Symposia*

(a) The 11th International History of Astronautics Symposium consisting of one session, organized by E. M. Emme (U.S.A.) and V. N. Sokolsky (U.S.S.R.), took place in Prague on 28 September 1977. The session was chaired by V. N. Sokolsky and G. S. James (U.S.A.), and comprised the seven papers listed in Annex I, para. 2.

(b) The 12th International History of Astronautics Symposium will be held in Dubrovnik in October 1978. Organized by E. M. Emme, V. N. Sokolsky and G. S. James, the programme includes ten papers relating to events in five different countries: F.R.G., Hungary, Sweden, U.S.A. and U.S.S.R.

3. *International Symposia on Space Economics and Benefits*

(a) The 7th International Symposium relating to the cost of space operations consisted of two sessions held in Prague on 26 and 27 September 1977, under the heading "Cost Effectiveness in Space Operations". The general Chairman was B. D. Haber (U.S.A.). The first session, organized and chaired by K. A. Ehricke (U.S.A.), was devoted to Socio-Economic Benefits of Space Operations. The second session dealt with Cost Reduction in Space Operations; organized by H. H. Koelle (F.R.G.), it was chaired by D. E. Koelle (F.R.G.). The papers presented in these two sessions are listed in Annex I, para. 3.

(b) The next Symposium is being organized under the title: "Space Economics and Benefits", which is also the new name of the former Committee on Cost Reduction in Space Operations. The Symposium will be held in Dubrovnik, under the general chairmanship of B. D. Haber. One session organized by D. E. Koelle will be on Cost Reduction in Space Operations, and the other, organized by K. A. Ehricke, on Socio-Economic Benefits of Space Operations. On the programme are papers by authors in seven countries: France, F.R.G., Indonesia, Switzerland, U.K., U.S.A. and U.S.S.R.

4. *International Review Meetings on Communication with Extra-Terrestrial Intelligence (CETI)*

(a) The two sessions of the 6th Review Meeting were held on 27 September 1977 in Prague, and were chaired by J. Billingham (U.S.A.) and R. Pesek (Czechoslovakia). The papers presented at this Meeting are given in Annex I, para. 4.

(b) The 7th International Review Meeting on Communication with Extra-Terrestrial Intelligence, also organized by J. Billingham and R. Pesek, will be held in Dubrovnik next October, with a programme consisting of papers from authors in six countries: Czechoslovakia, Hungary, Poland, U.S.A., U.S.S.R. and Yugoslavia.

5. *International Space Relativity Symposia*

(a) The Symposium, organized jointly by F. Estabrook (U.S.A.), Z. Horak (Czechoslovakia) and W. Wrigley (U.S.A.), was held in Prague on 30 September and 1 October 1977. Z. Horak chaired the two sessions during which eight papers were presented (see Annex I, para. 5.).

(b) The fifth International Space Relativity Symposium, organized by F. Estabrook and W. Wrigley, will be held during the 29th I.A.F. Congress in Dubrovnik in October. One session is planned with eight papers by authors in three countries: F.R.G., The Netherlands, and U.S.A.

6. *Round Tables on Scientific and Legal Aspects of International Cooperation in Remote Sensing*

(a) The Round Table held in Prague on 27 September 1977 was co-chaired by P. Contensou (France) and V. Kopal (Czechoslovakia). Participating in the discussion were seven invited lawyers and scientists: M. Bignier (France), M. Hirai (Japan), H. Disher (U.S.A.), L. Perek (United Nations), G. P. Zhukov (U.S.S.R.) (replacing D. Bordunov), E. Galloway (U.S.A.) and M. A. Ferrer (Argentina). Contributions by two authors who could not attend were read: G. Lafferanderie (ESA), and F. B. Andhyarujina (India). A report on the discussions at this Round Table is being published as an appendix to the Proceedings of the 20th International Colloquium on the Law of Outer Space of the International Institute of Space Law, held in Prague.

(b) In view of the interest expressed in this Round Table, a second Round Table on the same subject will be held in Dubrovnik in October 1978. This meeting is being organized by the same Chairmen, P. Contensou and V. Kopal. Scientists and lawyers from the U.N., ESA, France, U.S.A. and U.S.S.R. have been invited to participate in the Round Table.

7. *First International Symposium on Space Observations of Long-Term Climatic Changes*

Organized by M. Barrère (France) and S. S. Penner (U.S.A.), this Symposium consisted of a single session held in Prague on 30 September 1977. The papers presented in the Symposium are given in Annex I, para. 6. Edited by S. S. Penner, they are being published in Vol. V, No. 7/8 of *Acta Astronautica.*

8. *International Colloquia on Gasdynamics of Explosions and Reactive Systems*

(a) The 6th Colloquium was held as planned from 22 to 26 August 1977 in Stockholm (Sweden). I attended the meeting as President of the I.A.A. which co-sponsored it. The programme included sessions on the following topics: detonations (gaseous phase, condensed phase, heterogenous systems); explosions; turbulent combustion; rock fragmentation; diagnostics, lasers; shock waves; analysis; reactive systems. More than 100 papers and over 200 participants from 21 countries participated in the Colloquium.

(b) The 7th International Colloquium on Gasdynamics of Explosions and Reactive Systems is scheduled to be held in Göttingen (F.R.G.) from 20 to 24 August 1979. The programme co-Chairmen are N. Manson (France), A. K. Oppenheim (U.S.A.) and R. I. Soloukhin (U.S.S.R.). The local Organizing Committee is chaired by H. Gg. Wagner (F.R.G.).

4. Publications

1. *Acta Astronautica*

Volume IV and three double issues of Vol. V have appeared. Vol. V No. 7/8 is ready for circulation at the beginning of September, and two issues, No. 9 and No. 10, will follow within the next few weeks.

Volume IV, No. 5/6 contained selected papers of the 27th International Astronautical Congress, and Vol. V, No. 9 and No. 10 will contain papers of the 28th Congress. These special issues are edited by L. G. Napolitano, the Editor-in-Chief of I.A.F. publications.

An improvement in the publications schedule and in the diversification of the contents of the *Acta* has been noted and an effort has been made with some success to publish more proceedings of Academy symposia (see below, para. 2, Proceedings).

News items and announcements for publication in the *Ex-Mundo Astronautico* Section of the journal have been provided as usual by the Secretariat of the Academy.

An in-depth study of the *Acta*, covering editorial policies and practices, publication procedures and financial implications, has been initiated with a view to encourage greater involvement of the members of the Academy in the production of the journal, both as contributors and subscribers. The modalities of a procedure that would achieve this objective have not yet been worked out, but the idea is taking shape and I hope that it can be implemented in 1978.

2. *Proceedings*

A selection of papers of the second and third Space Relativity Symposia (Lisbon, 1975 and Anaheim, 1976) was published in Vol. V, No. 1/2 of *Acta Astronautica.* They were edited by W. Wrigley, who will also edit the papers of the subsequent symposia for later publication.

Papers presented at the CETI Review Meetings of the past three years have been edited by J. Billingham and are scheduled to appear in Vol. VI, No. 1/2 of *Acta Astronautica*.

Selected papers of the 1976–78 symposia on Space Economics and Benefits (see para. III. B. 3 above) are scheduled to appear in May 1979 in a special issue of *Acta Astronautica*. The manuscripts will be edited by B. D. Haber and K. A. Ehricke.

The proceedings of the 9th and 10th Space Rescue and Safety symposia have been prepared for publication in *Acta Astronautica*, under the editorship of P. H. Bolger.

A selection of papers of the 6th International Colloquium on Gasdynamics of Explosions and Reactive Systems will also be published in *Acta Astronautica*.

The proceedings in English of the 3rd through 6th History symposia (1969–72) were published for the Academy by NASA. Edited by R. C. Hall, they appeared in 1977 in two volumes as NASA. Edited by R. C. Hall, they appeared in 1977 in two volumes as NASA Conference Publication 2014. They are available from the Superintendent of Documents, U.S. Government Printing Office, Washington, D.C., U.S.A. Unfortunately, publication of the remaining symposia in English has been held up because the support needed to pursue the editing and printing has not yet been found.

The Russian version is progressing. The proceedings of the 2nd and 3rd symposia should be out at the end of 1978. Work has been completed on the 4th and 5th symposia and has begun on the 6th and 7th. As it was agreed that the papers in the Russian and English versions should be identical, the preparation of the books in Russian depends partly on the availability of the English publication.

5. Academy Committees

1. *Space Economics and Benefits Committee*

Upon the recommendation made by the Committee at its meeting in Prague last September, the name of the Committee was changed from "Cost Reduction in Space Operations" to "Space Economics and Benefits", which indicates more clearly the scope of its activity. Two sessions were requested at the 29th Congress for the next Symposium of the Committee (see para. III. B. 3b).

A new member has been appointed to the Committee: D. S. A. Sarkisjan (U.S.S.R.).

The Committee will meet again in Dubrovnik in October.

2. *Gasdynamics of Explosions and Reactive Systems Committee*

The Committee met during the Colloquium held in Stockholm in August 1977 and again in Prague in September, where arrangements for the 7th International Colloquium on Gasdynamics of Explosions and Reactive Systems were discussed. The venue for the next Colloquium has since been changed from Minsk (U.S.S.R.) as originally proposed, to Göttingen (F.R.G.).

Another meeting of the Committee is scheduled to be held in Dubrovnik next October.

3. *Communication with Extra-Terrestrial Intelligence (CETI) Committee*

At its meeting in Prague in September 1977, the CETI Committee was concerned not only with the programme of its next Review Meeting, but also with the protection of radio frequency bands for use in CETI research and appropriate representation in connection with the 1979 World Administrative and Radio Conference of the International Telecommunication Union where this question will be examined. Prof. Pesek, Chairman of the CETI Committee, briefed Dr. S. Saito, a Vice-President of the I.A.F., who represented the Federation at a preparatory meeting of the International Radio Consultative Committee (CCIR) in Kyoto in June of this year.

The Committee requested two sessions for another CETI meeting during the 29th I.A.F. Congress (see para. III. B. 4 b).

Two new members were appointed to the Committee: Dr. N. S. Kardachev (U.S.S.R.) and Dr. R. E. Edelson (U.S.A.).

The Committee will reconvene in Dubrovnik in October.

4. *History of the Development of Rockets and Astronautics Committee*

At its meeting in Prague in September 1977 the History Committee drew up plans for its 12th Symposium for 1978. Two sessions were requested but only one could be made available for this Symposium at the 29th I.A.F. Congress (see para. III. B. 2 b).

Among other subjects discussed at the meeting was the exchange of information on research in the history of astronautics in different countries, which the Committee recommended be continued along with the preparation of information bulletins in English and Russian for distribution to members of the Committee. The Committee heard reports on some of the problems regarding the history of rocket technology and astronautics as reflected in national museums in various countries and suggested that information on this question be exchanged, too. It was further agreed to collect information on the training of scientific personnel in the history of rocketry and astronautics, a subject to be discussed at a later meeting.

Mr. G. S. James has been appointed Vice-Chairman of the Committee to assist the co-Chairmen, E. M. Emme and V. N. Sokolsky. Other appointments are underway.

A meeting of the Committee is scheduled to be held in Dubrovnik in October.

5. *Space Rescue and Safety Studies Committee*

The Committee met in Prague and reviewed its plans for the next Symposium.

The participation of the Committee in the joint IAF/IAA Working Group on Satellite-Aided Search and Rescue broadens the scope of its activity, as reflected in the programme of the meetings held in Prague last year and the programme organized for the next Symposium in Dubrovnik (see para. III. B. 1 b).

In December, the Chairman of the Committee, Mr. P. H. Bolger, asked to be relieved of this responsibility because of the pressure of his work as Director of Safety Affairs in the U.S. Department of Transportation. At my request, Mr. Bolger kindly consented to remain in office until the appointement of a new Chairman, which is under way.

The Committee is scheduled to meet again in Dubrovnik in October.

6. *Scientific-Legal Liaison Committee*

The Committee met in Prague and proposed that a second Round Table be held in 1978 on the scientific and legal aspects of international cooperation in remote sensing, in view of the success obtained by the Round Table in Prague. A session was requested during the 29th I.A.F. Congress (see para. III. B. 6).

The Committee will meet again in Dubrovnik in October.

7. *Energy and Space Committee*

A proposal to establish a committee to study the interaction between energy problems and space science and technology was accepted by the Board of Trustees and the following members were appointed:

Chairman:	L. G. Napolitano (Italy)	
Vice-Chairman:	S. N. B. Murthy (U.S.A.)	
Members:	M. Barrère (France)	J. P. Layton (U.S.A.)
	G. G. Chernyi (U.S.S.R.)	S. S. Penner (U.S.A.)
	R. Goulard (France)	A. La Rocca (Italy)
	A. Jaumotte (Belgium)	G. Seibert (ESA)

The objectives of this Committee place emphasis on future possibilities of space technology to aid in finding solutions to energy problems on Earth, their potential realization and long-range applications, and international cooperation in this area.

The Committee is investigating the possibility of organizing a specialized symposium sometime next year and will discuss the project at a meeting in Dubrovnik in October.

8. *Other Committees*

Although the *Space Relativity Committee* did not meet in Prague, time was requested for the fifth Symposium to be held during the 29th I.A.F. Congress in 1978 (see para. III. B. 5 b). A meeting of the Committee is scheduled to take place in Dubrovnik in October.

The *Man in Space Studies Committee* cancelled its meeting in Prague last year and is waiting to reconvene until more information can be assembled regarding the possibility of organizing another symposium similar to those held several years ago.

The *MARECEBO Committee* has also suspended meeting until plans for future activities have been formulated.

6. Other Academy Activities

During the year, I participated in the meeting of the I.A.F. Bureau that were held at Prague in September 1977 and in Paris in April 1978.

Our Vice-President, Professor L. G. Napolitano, represented the Academy at the COSPAR meetings in Innsbruck (Austria) in June. These included a Symposium on graviational physiology of which the Academy was a co-sponsor.

7. Financial Support of the Academy

The resources of the Academy continue to be derived from grants and donations, voluntary contributions from the membership of the Academy, royalties received for the Academy's journal, *Acta Astronautica*, and registration fees for meeting sponsored by the Academy. While I am very grateful for the support provided by these sources, I am sorry to say that they are insufficient to provide stable or near-stable income.

Annual contributions have been recieved from the following supporting donors: Th. von Kármán Memorial Foundation, Messerschmitt-Bölkow-Blohm GmbH, Radio-Equipements-Antarès, Hughes Aircraft Co., U.S.S.R. Academy of Sciences, Czechoslovak Academy of Sciences, ERNO, Communications Satellite Corporation. To these should be added the stipend from the Daniel and Florence Guggenheim Foundation for the Daniel and Florence Guggenheim International Astronautics Award. I expect others to be received before the end of the year.

Voluntary contributions from Academy members amounted to more than in previous years, partly due to two contributions of $1000 each received from two members, one at the end of 1977, the other this year.

The difficult financial situation obliges the Academy to operate at minimal expense, and I thank our Secretary, Miss Hélène van Gelder, and her Assistant, Miss Michelle Pigé, for having carried out the work of the Secretariat in the usual satisfactory manner despite the restrictions.

8. Acknowledgements

I want to express my appreciation of the assistance and encouragement I received during the years from the members of the Board of Trustees and from the many Academy members and others who have contributed to our activities, financially or by their cooperation and services.

ANNEX 1

Scientific Meetings—Programme Details

1. *Tenth International Space Rescue and Safety Symposium*—30/9 and 1/10/1977 (See Report, para. III. B. 1 a).

"Space Shuttle Approach and Landing Test Project" by D. K. Slayton, T. McElmurry, NASA Johnson Space Center, Houston, Texas, U.S.A.

"The Synthesis of Control Algorithm ofr Maneuvering Space Shuttle Vehicle" by A. A. Zhevnin, Intercosmos Council, U.S.S.R. Academy of Sciences, Moscow, U.S.S.R.

"Survey of Space Flight Safety Systems—Seventh Supplement" by N. E. Brown, McDonnell Douglas Corporation, Seabrook, Texas, U.S.A.

"Designing Payloads for Safety on the Space Transportation System" by G. S. Canetti, Rockwell International, Downey, Calif., U.S.A.

"Systems Safety In Manned Space Flight" by J. B. Hammock, NASA Johnson Space Center, Houston, Texas, U.S.A.

"Shuttle Remote Manipulator System Safety and Rescue Support Capabilities" by G. D. Whitehead, Spar Aerospace Products Ltd., Toronto, Ont., Canada, J. W. Brown, NASA Johnson Space Center, Houston, Texas, U.S.A.

"Radiation Protection for Manned Orbital Stations" by N. N. Gurovsky, E. E. Kovalev, V. M. Petrov, Intercosmos Council, U.S.S.R. Academy of Sciences, Moscow, U.S.S.R.

"A Space-Based Public Service Platform for Terrestrial Rescue Operations" by R. Fleisig, Grumman Aerospace Corp., Bathpage, N.Y., U.S.A.

"The Probability of Decompression Sickness as a Result of Going Out in a Pressure Suit from a

Spacecraft with an Atmosphere Close to That on Earth" by A. S. Barer, L. G. Golovkin, S. N. Filipenkov, I. N. Chernyakov, A. A. Sheikin, Intercosmos Council, U.S.S.R. Academy of Sciences, Moscow, U.S.S.R.

"Advanced Vehicle Concepts for Earth-Orbit Transportation and Rescue" by R. C. Haefeli, Martin Marietta Aerospace, Denver, Col., U.S.A.

"Survey of Recovery Capabilities—1977 Annual Supplement—International Disastor Response—Space Systems and Terrestrial Elements" by G. W. Heath, SAR-ASSIST, Inc., Greenwich, Conn., U.S.A.

2. *Eleventh International History of Astronautics Symposium*—28/9/1977 (See Report, para. III. B. 2 a)

"Kepler's 'The Dream, or Lunar Astronomy' As a Predecessor of Space Research" by Z. Horský, Astronomical Institute of the Czechoslovak Academy of Sciences, Prague, Czechoslovakia

"First U.S.S.R. Rocket Experiments for Short-Wave Solar Radiation Research" by G. S. Ivanov-Cholodny, L. A. Vedeshin, A. B. Yakovleva, Intercosmos Council, U.S.S.R. Academy of Sciences, Moscow, U.S.S.R. (read by I. A. Merkulov)

"On Computer-Oriented Dynamic Modeling of Spacecraft: Historical Evolution of Eulerian Multi-body Formalism Since 1750" by R. E. Roberson, University of California, La Jolla, Calif., U.S.A.

"Organization and Results of Early Work at the U.S.S.R. Science Centre for Rocketry" by I. A. Merkulov, Intercosmos Council, U.S.S.R. Academy of Sciences, Moscow, U.S.S.R.

"Albert Fono: A Pioneer of Jet Propulsion, 1991–1972" by I. G. Nagy, Budapest, Hungary

"S. P. Korolev and Soviet Rocketry" by B. V. Rauschenbach, Institute of Control Sciences, Moscow, U.S.S.R. (read by I. A. Merkulov)

"Genesis of Liquid Hydrogen Propulsion Through 1945" by J. L. Sloop, Bethesda, Md., U.S.A. (read by G. S. James)

3. *Seventh International Symposium on Cost Effectiveness in Space Operations*—26 and 27/9/1977 (See Report, para. III B. 3 a)

"Evaluating the Benefits of Government Funded Research and Development Aimed at the Private Sector" by J. S. Greenberg, ECON, Inc., Princeton, N.J., U.S.A.

"Oceanautics—Space Technology and Undersea Research" by K. R. Stehling, US Department of Commerce, NOAA, Rockyville, Md., U.S.A.

"Economic Problems of Effective Application of Space Techniques for Studying Earth's Natural Resources" by S. A. Sarkisjan, S. S. Korunov, A. G. Gurov, I. I. Samoilova, Intercosmos Council, U.S.S.R. Academy of Sciences, Moscow, U.S.S.R.

"Space Solar Power Versus Space Communications" by S. K. Sarkar, General Directorate of Swiss PTT, Bern, Switzerland

"Socio-Economic Determinants of a Program for Lunar Industrialization in Support of Space Light Development Lunetta and Soletta" by K. A. Ehriche, Space International Corp., Downey, Calif., U.S.A.

"Cultural Engineering in Space" by S. C. Eilenberg, S. L. Eilenberg, Rockwell International, Downey, Calif., U.S.A.

"A Scenario for the Planning, Development and Integration of a Payload in the Space Transportation Systems" by G. F. Dowdall, G. S. Canetti, Rockwell International, Calif., U.S.A.

"Organizing for Low-Cost Space Transportation" by C. Lee, NASA Headquarters, Washington, D.C., U.S.A.

"Future Low-Cost Space Transportation Systems Analysis—Part III: Preliminary Size Optimization" by D. E. Koelle, Messerschmitt-Bölkow-Blohm GmbH, Ottobrunn, H. H. Koelle, Technical University of Berlin, Berlin, F.R.G.

"Trade-Off Between Cost and Availability for Satellite Systems" by R. Chaufriasse, Centre National d'Etudes Spatiales, Toulouse, France

"Maintainability and Maintenance Optimization Techniques for Reusable Spacecraft" by J. Leupen, Amsterdam, The Netherlands

"Concepts and Procedures for Low-Cost Production of the Shuttle-External Tank" by G. E. Smith, Martin-Marietta Corporation, New Orleans, La., U.S.A.

"Economic Effectiveness of Rational Organization of Transportation Operations in Reusable Space Transport Systems" by S. A. Sarkisjan, S. S. Korunov, I. I. Samoilova, A. G. Gurov, Intercosmos Council, U.S.S.R. Academy of Sciences, Moscow, U.S.S.R.

4. *Sixth International Review Meeting on Communication with Extra-Terrestrial Intelligence (CETI)*—27/9/1977 (See Report, para. III. B. 4 a)

"The Design of Ground-Based Antennas for SETI" by S. von Hoerner, National Radio Astronomy Observatory, Green Bank, Va., U.S.A.

"Strategic Considerations in SETI and a Microwave Approach" by C. L. Seeger, NASA Ames Research Center, Moffett Field, Calif., U.S.A. (read by J. Billingham)

"An Experimental Protocol for a Search for Radio Signals of Extra-Terrestrial Origin in the Presence of Man-Made Radio Frequency Sources" by R. E. Edelson, Jet Propulsion Laboratory, Pasadena, Calif., U.S.A.

"On the Active and Passive CETI from the Earth Satellite Orbit" by M. Subotowicz *et al.*, Polish Astronautical Society, Warsaw, Poland

"Review of Contemporary Opinions on the Problem of the Origin of Life" by L. M. Muhin, Intercosmos Council, U.S.S.R. Academy of Sciences, Moscow, U.S.S.R.*

"To the Problem of the Definition of Intelligence" by J. M. Krein, Intercosmos Council, U.S.S.R. Academy of Sciences, Moscow, U.S.S.R.*

"On the Strategy of Searching for Extra-Terrestrial Civilization" by N. S. Kardashev, Intercosmos Council, U.S.S.R. Academy of Sciences, Moscow, U.S.S.R.*

"Natural Law for All Sentiment Beings" by E. Fasan, Neunkirchen, Austria

"Information Aspects of CETI Program" by M. Weisenverger, Charles University, Prague, Czechoslovakia

5. *Fourth International Symposium on Space Relativity*—30/9 and 1/10/1977 (See Report, para. III. B. 5 a)

"Geodesy Information in a Modified Relativity Mission with Two Counter-Orbiting Polar Satellites" by J. V. Breakwell, C. W. F. Everitt, D. Schaechter, R. A. van Patten, Stanford University, Stanford, Calif., U.S.A.

"On a Non-Eclipse Optical Determination of the Gravitational Deflection of Light by the Sun" by F. Handler, R. Matzner, University of Texas, Austin, Texas, U.S.A.

"Testing Gravity Theories by a Solar Probe" by E. A. Roth, European Space Agency, Darmstadt, F.R.G.

"Gravity Research at Cottrell Observatory" by V. S. Tuman, California State College, Stanislaus, Turlock, Calif., J. D. Anderson, E. L. Lau, Jet Propulsion Laboratory, Pasadena, Calif., R. Uhrhammer, University of California, Berkeley, Calif., U.S.A.

"Inertial Frames Determined from the Rotational Laws Observed in Spiral Galaxies" by Z. Horák, Technical University of Prague, Prague, Czechoslovakia

"Methods of Verification of the Time Dilatation by Means of Lasers" by J. Leupen, Amsterdam, The Netherlands

"The Results of the Gravitational Redshift Rocket Probe Experiment" by R. F. C. Vessot, M. W. Levine, Smithsonian Institution, Astrophysical Observatory, Cambridge, Mass., U.S.A.

"The Technical Feasibility of a General Relativity Experiment Extending to the Sun" by R. F. C. Vessot, M. W. Levine, K. Nordvedt, Jr., Smithsonian Institution, Astrophysical Observatory, Cambridge, Mass., U.S.A.

6. *First International Symposium on Space Observations of Long-Term Climatic Changes*—30/9/1977 (See Report, para. III. B. 7)

"Introductory Remarks: Space Observations of Long-Term Climatic Changes Produced by Escalating Energy Use" by S. S. Penner, University of California, San Diego, La Jolla, Calif., U.S.A.

"Energy Use and Climate Changes" by J. Williams, W. Häfele, International Institute for Applied Systems Analysis, Schloss Laxenburg, Austria

*Read with additional comments by I. Shklovski

"Monitoring of Thickness Changes of the Continental Ice Sheets by Satellite Altimetry" by R. L. Brooks, H. R. Stanley, Washington Analytical Services Center, Inc., Wolf Research & Development Group, Pocomoke City, Md., U.S.A.

"Space Observations of Long-Term Climate Changes" by L. Jaffe, NASA Headquarters, Washington, D.C., U.S.A.

"The Recent Temperature Trend in the Low Stratosphere" by J. K. Angell, Air Resources Laboratory, NOAA, Silver Spring, Md., U.S.A.

"The Effect of Energy Consumption on Climate" by I. Peychès, France (presented by E. A. Brun).

"New Trends in the Development of Meteorological Satellites" by A. Villevieille, France (presented by E. A. Brun)

Acta Astronautica. Vol. 6, pp. 13–16. Pergamon Press 1979. Printed in Great Britain

Ex Mundo Astronautico

I.A.F. Membership List
1978–1979

(V.M.) = Voting Member
(M.) = Member
(I.M.) = Institution Member

Argentina

(V.M.) Asociación Argentina de Ciencias Espaciales (AACE)
Perú 759
Buenos Aires

(M.) Centro de Investigaciones Espaciales
Venezuela 4383
Buenos Aires 1211

Australia

(V.M.) Astronautical Society of Western Australia
Box E 254, G.P.O.
Perth 6001

(M.) The Astronautical Society of Australia, Ltd.
P.O. Box 93
Black Rock, 3193

Austria

(I.M.) Austrian Solar and Space Agency—ASSA
Garnisongasse 7
A–1090 Vienna

Belgium

(V.M.) Association Belge des Ingénieurs et Techniciens de l'Aéronautique et de l'Astronautique—ABITA
c/o Direction Technique SABENA
Aéroport Bruxelles-National
1930 Zaventem

Bulgaria

(V.M.) Bulgarian Astronautical Society
Union of Scientific Workers
18 Tolbouhine Boulevard
Sofia

Canada

(V.M.) Canadian Aeronautics and Space Institute—CASI
Saxe Building, 60
75 Sparks Street
Ottawa, Ont. K1P 5A5

China (Republic of)

(V.M.) Aeronautical and Astronautical Society of the Republic of China
41 Nan Hai Road
Taipei, Taiwan (Formosa)

Cyprus

(V.M.) Cyprus Astronautical Society
P.O. Box 210
Limassol

Czechoslovakia

(V.M.) Commission on Astronautics
Czechoslovak Academy of Sciences
Národní 3
111 42 Prague 1

Denmark

(V.M.) Dansk Astronautisk Forening
Postbok 31
1002 Copenhagen K

Finland

(V.M.) Suomen Avaruustutkimusseura
Postilokero 507
SF-00101 Helsinki 10

France

(V.M.) Association Aéronautique et Astronautique de France—AAAF
80 Rue Lauriston
75116 Paris

(M.) Association pour l'Etude et la Recherche Astronautique et Cosmique—AERA
Boîte Postale 221-01 RP
75017 Paris

(M.) Société Chimique de France
250 Rue Saint-Jacques
75005 Paris

German Democratic Republic

(V.M.) Astronautische Gesellschaft der DDR
Poststrasse 4–5
102 Berlin

German Federal Republic

(V.M.) Deutsche Gesellschaft für Luft und Raumfahrt e.V.—DGLR
Goethestrasse 10
5 Köln-Marienburg

(M.) Hermann-Oberth-Gesellschaft
Flughafendamm (Raum 40)
D-2800 Bremen 1

Greece

(V.M.) Hellenic Astronautical Society
Voulis Street 14
Athens 126

(M.) Scientific Group for Space Research
Astronomical Department
13 Meliteon Street
Athens 305

Hungary

(V.M.) Magyar Asztronautikai Egyesület
Kossuth Lajos tér 6/8
Budapest

India

(V.M.) Indian Rocket Society
c/o Vikram Sarabhai Space Center
ISRO P.O.
Trivandrum 695022

Indonesia

(V.M.) Indonesian Space Society
(Perkumpulan Antariksa Indonesia)
c/o LAPAN
Jl. Pemuda Persil N° 1
Jakarta IV

Iran

(V.M.) Iranian Astronautical Society
Institute of Geophysics
University of Teheran
Amirabad, Teheran

Israel

(V.M.) Israel Society of Aeronautics and Astronautics
Department of Aeronautical Engineering
Technion City, P.O. Box 4910
Haifa

Italy

(V.M.) Associazione Italiana di Aeronautica e Astronautica—AIdAA
Via Po, 50
00198 Rome

(M.) Associazione Italiana di Medicina Aeronautica e Spaziale
Via P. Gobetti 2A
00185 Rome

(M.) Istituto Italiano di Diritto Spaziale
Via Giulia 251
00186 Rome

(M.) Istituto Italiano di Navigazione
Piazza Cavour, 25
00193 Rome

(I.M.) Consiglio Nazionale delle Ricerche
Piazzale delle Scienze, 7
00100 Rome

Japan

(V.M.) Japanese Rocket Society
Business Center for Academic Societies Japan
4–16, Yayoi 2-chome, Bunkyo-ku
Tokyo

(M.) Japan Astronautical Society
(International Section)
4–23–8 Nakaochiai, Shinjuku-ku
Tokyo

(I.M.) Space Activities Promotion Council, Keidanren
9–4, Ohtemachi 1-chome, Chiyoda-ku
Tokyo 100

Liechtenstein

(V.M.) Förderkreis Hermann Oberth im Fürstentum Liechtenstein
P.O. Box 200
FL-9494 Schaan

Mexico

(V.M.) Sociedad Mexicana de Estudios Interplanetarios
Apartado 44009
Mexico 12, D.F.

Netherlands

(V.M.) Nederlandse Vereniging voor Ruimtevaart
c/o National Aerospace Laboratory NLR
31 Voorsterweg
8300 AD-Emmeloord

(I.M.) Netherlands Agency for Aerospace Programs
Kluyverweg 1—P.O. Box 35
Delft 2209

Norway

(V.M.) Norsk Astronautisk Forening
Postboks 43
Blindern, Oslo

Poland

(V.M.) Polskie Towarzystwo Astronautyczne
P.K.i.N.p. 23–23
00–901 Warsaw

Portugal

(V.M.) Grupo Portugués de Astronáutica
Instituto Superior Técnico
Avenida Dr. A. José de Almeida
Lisbon

Rumania

(V.M.) Commission d'Astronautique de l'Académie de la République Populaire Roumaine
15 Rue Constantin Mille
Bucarest

Spain

(V.M.) Agrupación Astronáutica Española
Rosellón 134, entlo. 2°
Barcelona 11

(M.) Asociación Española Astronáutica
Paseo del Pintor Rosales, 34
Madrid 8

Sweden

(V.M.) Swedish Society of Aeronautics and Astronautics
Box 11021
S-161 11 Bromma

Switzerland

(V.M.) Swiss Association for Space Technology—SAST
Postfach 2613
CH-3001 Bern

(M.) Swiss Association for Rocketry—SAFR
Lidostrasse 5
CH-6006 Luzern

(I.M.) Swiss Transport Museum
Lidostrasse 5
CH-6006 Luzern

United Kingdom

(V.M.) British Interplanetary Society
12 Bessborough Gardens
London S.W.1V 2JJ .

(M.) Association in Scotland to Research into Astronautics—ASTRA
57 Gordon Road
Hamilton ML3 9LG, Lanarkshire
Scotland

(M.) Royal Aeronautical Society
Astronautics & Guided Flight Section
4 Hamilton Place
London W.1

U.S.S.R.

(V.M.) Intercosmos Council
U.S.S.R. Academy of Sciences
Leninsky Prospect 14
117901 Moscow V-71

U.S.A.

(V.M.) American Institute of Aeronautics and Astronautics—AIAA
1290 Avenue of the Americas
New York, N.Y. 10019

(M.) American Astronautical Society
6060 Duke Street
Alexandria, Va. 22304

(M.) Aerospace Medical Association
Washington National Airport
Washington, D.C. 20001

(I.M.) Rocket Research Institute, Inc.
3262 Castera Avenue
Glendale, Calif. 91208

Yugoslavia

(V.M.) Jugoslovensko Astronauticke i
Raketno Drustvo
Bulevar Revolucije 44
11000 Belgrade

(M.) Yugoslav Aerospace Society
27 Marta Street 80
Faculty of Mechanical Engineering
11000 Belgrade

Acta Astronautica. Vol. 6, pp. 17–19. Pergamon Press 1979. Printed in Great Britain

Ex Mundo Astronautico

In Memoriam

A. Val Cleaver
1917–77

Val Cleaver died after some illness, nevertheless very unexpectedly, in London on September 16, 1977.

His professional career is clear and straightforward: after fighting for his desired education—engineering—he joined deHavilland Aircraft Corporation, and worked in the field of propellers. In 1946, he became Chief Engineer—and changed to propulsion. There, he found his life's vocation: modern liquid propellant rocket engines. When the British Aircraft industry was restructured, he went with the job, and was appointed Rolls Royce's Director of the Rocket Engines Department. A major product was the Blue Streak—later first stage of the ill-fated Europe Space Carrier Vehicle—propulsion system. Blue Streak's flight history is well known: perfection.

Less known: from his work came the rocket engine for a turbojet/rocket fighter aircraft of the F 104 class—well before this bird was available. "Project cancelled", like so many European post-war activities.

His mind and his love belonged to astronautics. He saw his professional work mainly in providing his contribution to the giant task ahead. Being eminently practical-minded, he soon joined the British Interplanetary Society, and became its Chairman soon after WW II. Later, he was elected honorary member of each of the more important astronautical societies. And—I am inclined to say: of course—he was a founding member of the International Academy of Astronautics.

His publications are numerous. I shall mention only two, which I believe to be most important: in Sept. 1949, the BIS Journal printed "The Atomic Rocket", which he had written together with Les Shepherd. An in Jan. 1954 "A Programme for Achieving Interplanetary Flight": for the first time, reasonable cost and time schedules were given.

Those two publications portray the man: daring and forwardlooking, and yet very practical and sound in judgement. No blue-eyed dreamer was he.—Dreams, yes, but tempered by common sense and experienced judgement. The latter qualities seem to have been lost in several recent proposals of contemporaries. We will miss Val, who saw not only the goal, but also the way leading there, with all its obstacles.

Nobody can write about Val Cleaver without mentioning his deep humanitarian spirit, his frankness in technical debate and his firm European roots. The world is poorer for his loss. Within the small world of European astronautics, the gap cannot be closed—as a mutual friend said to me about Val: "they don't make them like him any more".

H. O. Ruppe

Fernand Vinsonneau
1902–78

Membre de l'Académie Internationale d'Astronautique depuis 1965, Monsieur Fernand Vinsonneau vient de mourir à Cannes, au mois de mai dernier, à l'âge de 76 ans.

Ancient élève de l'Ecole Polytechnique de Paris, il devait faire carrière principalement dans l'industrie aéronautique: S.N.C.A.S.O., puis Sud-Aviation d'où il avait été détaché en 1959 à la S.E.R.E.B., Société créée principalement pour piloter la réalisation de la force de frappe balistique française mais aussi des lanceurs spatiaux. C'est là que commence la carrière spatiale de Monsieur Vinsonneau en liaison avec la réalisation du lanceur Diamant. Il occupe à la S.E.R.E.B. les fonctions de Directeur Technique jusqu'en Septembre 1961, date à laquelle il est nommé Secrétaire Général d'EUROSPACE qui vient d'être fondé et dont il sera le premier Secrétaire Général.

C'est donc lui qui a effectivement mis sur pied, organisé et donné son orientation initiale à l'Association *Eurospace* qui attira rapidement l'attention des milieux gouvernementaux et scientifiques par ses "Propositions pour un programme spatial européen" publiées en 1963, puis par sa proposition d'un "Transporteur Aérospatial" réutilisable qui, dès 1964, frayait—sous les quolibets des Américains soucieux alors de "vendre" leurs fusées consommables—la voie du "Space Shuttle".

Dès les premiers travaux d'*Eurospace*, Monsieur Vinsonneau avait apporté une attention particulière à la promotion de programmes européens de satellites de télécommunications géostationnaires. La première étude du marché ouvert à de tels satellites fut menée en 1967 sous sa direction par *Eurospace* avec Metra-International comme consultant.

Monsieur Vinsonneau a quitté *Eurospace* pour prendre sa retraite en Octobre 1967, mais l'on sait par la participation active qu'il prit à l'Académie que l'Espace continua d'être pour lui jusqu'à sa mort son "violon d'Ingres" préféré.

J. Delorme

Henri Moureu
1899–1978

M. Henri Moreu, membre de l'Académie des Sciences de l'Institut de France, né â Paris le 2 août 1899, est mort à Pau le 13 juillet 1978.

Les brillants résultats de ses recherches en chimie organique et en chimie minérale ont été obtenus d'une part au Collège de France où il a occupé successivement des postes d'Assistant et de Sous-Directeur (1924–41), d'autre part au Laboratoire Municipal de Paris à la direction de laquelle il a été appelé dès 1941.

Nous ne ferons que rappeler ici ses travaux sur les fusées et la propulsion par réaction qui l'ont conduit, à la demande de l'Etat-Major de l'Armée, a créer en 1945 le Centre d'Etudes des Projectiles Autopropulsés (CEPA); il en assura constamment la direction scientifique. Il avait, dès avant cette époque, la prescience de l'avenir de la fusée; en 1947, lors de la "Semaine Nationale de Sécurité", tenue à Paris, il a précisé ses vues dans un article ayant pour titre "Les engins stratégiques à longue portée, forme nouvelle de la guerre totale".

L'importance des recherches du CEPA, notamment celles concernant les propergols ou la stabilisation des systèmes linéaires, ont justifié la place prise par Henri Moureu dans l'astronautique. Il a assumé avec un grand dévouement sa présidence à la Société Française d'Astronautique. Membre de l'Académie Internationale d'Astronautique, il a assisté à plusieurs des congrès internationaux d'astronautique. Il a facilité l'installation à la rue Saint-Jacques du Secrétariat de la Fédération Internationale d'Astronautique.

E. A. Brun

Harry Ross

1904–78

Harry Ross who died on 8 May 1978 at the age of 73 was among the pioneer band of contributors to the concepts, practice and philosophy of astronautics. He joined the British Interplanetary Society in 1937 and participated in the BIS Lunar Spaceship study of 1937–39. In 1946 he joined forces with another early BIS member, the late Ralph A. Smith, to revise the pre-War design of the Lunar-landing vehicle in the light of war-time developments in liquid-fuel propulsion. Subsequently they worked out designs for a man-carrying sub-orbital rocket based on a modification of the German V-2, a scheme very similar to that adopted by the United States in 1961 using the RedstoneMercury rocket. Together they formulated a design for a manned space-station powered by solar energy, which formed the basis of a paper "Orbital Bases" published in 1949 and outlining a method for landing on the Moon using a space vehicle refuelled by unmanned telecontrolled "tankers" in lunar orbit. In 1950 he published a detailed design of a spacesuit for use on the Moon. In 1967 Mr. Ross was elected a Corresponding Member of the International Academy of Astronautics in the Engineering Sciences Section and in 1971 he received the BIS Bronze Medal for outstanding contributions to astronautics. During the last ten years, in spite of failing health, he developed his philosophy of astronautics in a series of articles, the final one "Quo Vadis" appearing in the BIS publication *Spaceflight* in July 1978. In his lifetime he saw the practical realization of many of the astronautical concepts and designs to which he had contributed 30 or more years previously and increasingly expounded on the role of Space in the shaping of human destiny.

G. V. GROVES

ANDRÉ LALLEMAND

1904–1978

André Lallemand, membre correspondant de l'Académie Internationale d'Astronautique, s'est éteint au printemps 1978. Né le 29 Septembre 1904 à Cirey, dans la Côte d'Or en France, André Lallemand devint agrégé de Physique en 1927, docteur en Sciences en 1934, professeur au Collège de France à Paris en 1961. Il commença sa carrière d'Astronome à l'Observatoire de Strasbourg, puis fut nommé en 1942 à l'Observatoire de Paris. L'oeuvre maîtresse de André Lallemand, qui le conduisit à l'Académie des Sciences en 1961, reste la mise au point de la caméra électronique. Pionnier incontesté dans la discipline maintenant appelée "électronographie", il réalise la première image stellaire amplifiée par transformation photoélectrique des photons en électrons, lesquels sont accélérés et focalisés sur une émulsion photographique. Dans la suite, André Lallemand développe la "caméra électronique" pour en faire l'outil de base de l'astrophysique. Ces instruments équipent de nombreux télescopes. Dans le sillage de ces recherches, André Lallemend développe aussi les photomultiplicateurs qui resteront longtemps les plus sensibles et les mieux adaptés aux besoins de l'Astrophysique.

Ce sont les procédés électronographiques de Lallemand qui ont été utilisés pour réaliser la première caméra électronique ultraviolette posée sur le sol de la Lune. Ce sont aussi les connaissances que André Lallemand rend disponibles à l'industrie qui conduisirent à la réalisation aux Etats-Unis des premiers photomultiplicateurs spatiaux sélectionnés pour les vols dans l'espace.

AUDOUIN DOLLFUS